Thermal Processing of Municipal Solid Waste for Resource and Energy Recovery

Thermal Processing of Municipal Solid Waste for Resource and Energy Recovery

Norman J. Weinstein
President

Richard F. Toro
Vice President

RECON SYSTEMS, INC.
Princeton, New Jersey

P. O. Box 1425, Ann Arbor, Michigan 48106

Library of Congress Catalog Card No. 76-1721
ISBN 0-250-40127-4

Manufactured in the United States of America

PREFACE

Until recent years, incineration has been considered an expensive alternative to landfilling for disposal of municipal solid waste, to be used only when landfill sites were not readily available. Today, thermal processes, which include incineration, pyrolysis, and combined refuse/fossil fuel combustion, with energy and/or resource recovery can be considered as competitors for landfilling, both because of the increased cost of landfilling performed in an environmentally sound manner and because of the increased value of energy, metals and glass which can be extracted from municipal solid waste.

However, thermal processing of solid wastes and associated resource recovery systems are in a state of transition from developmental to fully operational stages. The performance of waterwall incinerators which generate steam has been fully proved both in Europe and North America. However, no waterwall incinerator projects in the U.S. are yet on sound footing with regard to external steam sales. Three modern waterwall incinerators in the U.S. are simply condensing all the steam generated, except for the small amounts being used internally. A new waterwall incinerator was recently started up and will supply steam to 27 downtown buildings for heat and air conditioning. Hopefully this project will succeed. Because of imbalance between supply and demand for energy, only projects which have been planned with extreme care will fully meet expectations.

On the other hand, combined refuse/fossil fuel firing in existing steam boilers makes use of already available facilities for energy distribution and already available markets. The uncertainties which do exist are mainly technical. Although a large-scale test has been conducted on a refuse/coal fired boiler, each new test with varying boiler designs, other types of coal and oil firing raise technical questions of corrosion, erosion, fuel handling and air pollution control. Careful engineering will be required to ensure broad success of this promising approach to thermal processing with energy recovery.

Several pyrolysis processes are available for converting municipal solid waste to fuels. One plant, in an early stage of operation, converts the fuel to steam; another is under construction; a third has been undergoing tests in a large-scale prototype. While the fuels produced are not conventional in the sense of fuels widely used today, they should find ready outlets under contract. The importance of good project and contract planning holds as true for pyrolysis processes as for steam-generating incinerators.

Resource recovery can enhance any of the thermal processes discussed above. Rapidly emerging technology should eventually allow recovery not only of ferrous metals but also color-sorted glass, aluminum, other metals, and even paper fiber. Some resource recovery plants may separate a combustible fraction which can be transported elsewhere for use in pyrolysis or combustion with energy recovery.

Some of the steps in resource recovery can be considered proven, for example, preparation of a combustible fraction and ferrous metal recovery from either mixed refuse or from incinerator residue. However, no complete system recovering the full range of potentially valuable energy and materials is yet fully operational and economical to the point of paying for solid waste disposal. Even with technical success, many marketing and end use problems will remain for some time to come, as industry and others become accustomed to unfamiliar materials and energy forms available in municipal solid waste.

Since the success of thermal processing hinges on efficient environmental control, Chapters 6 and 7 discuss air and water pollution control and residue disposal. The data presented are primarily for incinerators, but the principles discussed hold for other thermal processes as well.

The terms "municipal solid waste," "solid waste," and "refuse" are used interchangeably in this publication. No special significance should be ascribed to the use of a specific term unless it is further modified, *e.g.,* "prepared refuse" or "shredded solid waste."

Norman J. Weinstein
Richard F. Toro
December 1975

ACKNOWLEDGMENTS

Many people have contributed to the concept and contents of this publication. An earlier publication, *Municipal-Scale Incinerator Design and Operation* (1969), by Jack Demarco, Daniel J. Keller, Jerold Leckman and James L. Lewton provided the basis for initiating the work. Other U.S. Environmental Protection Agency personnel who participated at various stages include William C. Achinger, Edward L. Higgins, Harvey Rogers, and Steven Hitte, the Project Officer.

Much of the work was done by RECON SYSTEMS personnel under EPA Contract No. 68-03-0293. Major contributions were made by Arthur T. Goding, Jr., Charanjit Rai and Robert M. Wolfertz. A special note of thanks is due to Mrs. Gladys Freeland and Mrs. Carol Picker.

Thanks are also due to the many incinerator operators and others active in thermal processing who gave of their time to provide RECON SYSTEMS with maximum insight into this aspect of municipal solid waste management.

To Ann and Nancy

CONTENTS

Chapter

LIST OF FIGURES

Figure

LIST OF TABLES

Table

Table

Table

Table

CONVERSION FACTORS

B x 42 = U.S. gal

BTU x 252 = calories

°C = (°F - 32) ÷ 1.8

$/MT x 0.907 = $/ST

°F = (°C x 1.8) + 32

ft x 0.3048 = meters

cal x 3.968 x 10^{-3} = BTU

cal/g x 1.80 = BTU/lb

CF x 0.02832 = CM

CM x 35.31 = CF

grains/SCF x 2.29 = grams/SCM

grams x 15.43 = grains

in. x 2.54 = cm

in. H_2O x 1.868 = mm Hg (pressure differential)

Kcal x 3.968 = BTU

kg x 2.2046 = lb

lb x 7000 = grains

lb x 0.454 = kg

lb/CF x 0.01602 = g/cc

lb/CF x 27 = lb/CY

lb/CY x 1.308 = lb/CM

MT x 1000 = kg

MT x 2205 = lb

MT x 1.1025 = ST

SCF (60°F, 1 atm) x 0.0268 = NCM (0°C, 1 atm)

SCF (70°F, 1 atm) x 0.0263 = NCM (0°C, 1 atm)

SCF (70°F, 1 atm) x 0.0283 = SCM (70°F, 1 atm)

ST x 2000 = lb

ST x 0.9070 = MT

(ST/24 hr) x 0.0378 = MT/hr

60°F = 15.6°C

70°F = 21.1°C

psia ÷ 14.7 = atmospheres absolute

$\frac{\text{psig} + 14.7}{14.7}$ = atmospheres absolute

TABLE OF ABBREVIATIONS

ACFM = actual cubic feet per minute
bbl = barrel(s) (42 U.S. gal)
cal = calories
cc = cubic centimeter(s)
CF = cubic feet
CFH = cubic feet per hour
CFM = cubic feet per minute
cm = centimeter(s)
CM = cubic meter(s)
CY = cubic yard(s)
$M = thousands of dollars
$MM = millions of dollars
ft = feet
ft^3 = cubic foot
gal = gallon(s) (U.S.)
g = gram(s)
GPM = gallons (U.S.) per minute
gr = grains
hr = hour(s)
in. = inch(es)
Kcal = kilocalories
kg = kilogram(s)
Kw = kilowatt (s)
lb = pound(s)
m^3 = cubic meter
min = minute(s)
ml = milliliter(s)
mm = millimeter(s)
MT = metric ton(s)
MT/hr = metric tons per hour
NCM = normal cubic meter(s) (0°C, 1 atm or 70°F where indicated)
SCF = standard cubic feet (60°F, 1 atm; or 70°F where indicated)
SCM = standard cubic meters (70°F, 1 atm)
ST = short tons
ST/day or ST/D = short tons per 24-hour day
tons/ton or tons per ton always refers to consistent weight units, for example MT/MT, ST/ST, Kg/Kg, lb/lb
TPD = short tons/day
vol = volume
WG = water gage or water column (pressure differential)
wt = weight
yr = year(s)

CHAPTER 1

THERMAL PROCESSING OF SOLID WASTES: AN INTRODUCTION

A municipal solid waste management system consists of a number of steps, starting with collection and continuing through transportation, pretreatment, treatment, environmental controls, disposal of residues and recycle of by-products. This publication deals with municipal solid wastes from the point of delivery to a thermal processing plant to the discharge of residues and recycle of by-products from the process. The following introductory discussion highlights the heart of the system, the thermal processing step.

The primary objective of any effective waste management system is disposal, while avoiding or minimizing damage to the environment. This objective may be met by reclaiming useful materials and/or conversion of waste components to benign or useful materials. For example, in the sanitary landfill of municipal solid wastes, the wastes are dumped onto specially chosen and prepared sites, compacted, isolated into cells using a cover such as earth, and allowed to slowly change by biochemical action to a complex, but hopefully benign, material allowing subsequent use of the land site.

Thermal processing of solid wastes is the elevated temperature treatment of those wastes in suitably designed equipment so as to convert the waste components into benign or useful materials. In practice, thermal processing may be accomplished in the presence of substantial quantities of added oxygen (usually air) in a process called incineration; or thermal processing may be accomplished with little or no oxygen added to that already chemically bound within the waste.

Historically, the only thermal process of importance with respect to municipal solid waste has been incineration. In an effective incineration process, the combustion gases are composed almost entirely of carbon dioxide, water, nitrogen and oxygen, all of which are normal atmospheric constituents. The residue should contain little or no combustible material.

In recent years, other thermal processes have been developed to produce products which are useful as fuels, and possibly as chemical raw materials. These processes can be subdivided into three categories: simple pyrolysis, where little or no oxygen is added to the thermal treatment zone; partial oxidation, where appreciable amounts of oxygen or air are added, producing substantial quantities of carbon monoxide and hydrogen in addition to carbon dioxide and water; and reduction, where either hydrogen or carbon monoxide is reacted with solid waste. All of these categories are usually referred to as pyrolysis, and this practice will be followed here.

This publication will deal mainly with modern evolving incineration processes and pyrolysis processes in which resource and energy recovery is practiced. The selection of a suitable process and some of the highlights of the major types of thermal processes which are commercially available are discussed in the following paragraphs.

SELECTION OF A SUITABLE PROCESS

In the current social and regulatory environment, the choice among alternative solid waste disposal processes will normally be made by considering impact on local and regional environment, local and regional optimum land use, net processing costs, and other specific local problems. For example, one should consider the desirability of dealing with secondary sewage sludge, special industrial and commercial wastes, and special administrative and operating problems. The choice may also be colored by the attitude of the local community toward environmental quality and resource conservation.

In the evolving technology of municipal solid waste disposal, the state of development of new processing techniques must also be a prime consideration in process selection. For example, in its current state of development (1975), the choice of a pyrolysis process would necessarily entail greater risks than the choice of a proven incineration process, though the risk might be worth taking because of potential economies or because of questions of resource conservation.

It is necessary to make a realistic up-to-date assessment of the status of all the options available in order to select the best process for the needs of a particular community. Recycling opportunities must be considered as an integral part of such an assessment. It is a prime purpose of this publication to acquaint those responsible for such decisions with the current state-of-the-art, and with important factors which should be considered in selecting a suitable municipal solid waste thermal processing system.

INCINERATION PROCESSES

Incineration has been the traditional competitor to landfill in areas where insufficient suitable landfill capacity is available within an economic haul distance. Although most incinerators built in the past could not meet today's performance criteria, a well-designed, carefully-operated incinerator reduces the weight and volume of municipal solid waste to produce a nuisance-free residue which can be used as a fill material. Table 1 compares volume reduction obtainable by thermal processes

Table 1. Calculation of Volume Reduction by Various Solid Waste Disposal Systems

	Original Volume as Fractions	Reduction Factor	Final Landfill Volume as Fraction as Original Volume
Sanitary Landfill[1]			
Incinerable Waste[a]	0.8	0.166	0.133
Bulky and Nonincinerable Waste	0.2	0.5	0.100
	1.0		0.233
Sanitary Landfill with Shredding and Resource Recovery			
Total Waste[a]	1.0	$(0.125^2)^b$	$(0.125)^b$
Conventional Incineration[1]			
Incinerable Waste[a]	0.8	0.0145	0.012
Bulky and Nonincinerable Waste	0.2	0.5	0.100
	1.0		0.112
Incineration with Shredding and Resource Recovery			
Total Waste[a]	1.0	$(0.008^2)^b$	$(0.008)^b$
Pyrolysis Processes with/without Resource Recovery[c]			
Total Waste[a]	1.0	$(0.004\text{-}0.03)^b$	$(0.004\text{-}0.03)^b$

[a] Bulk density assumed to be 89 kg/CM (150 lb/CY).

[b] Numbers in parentheses are somewhat speculative since little data are available for confirmation.

[c] Numbers for pyrolysis do not include ash content of liquid or solid fuels which will become residual product when the fuel is burned.

with that encountered in sanitary landfill. Gases discharged to the atmosphere are treated to meet governmental standards for emission of particulates and chemical constituents. Water used for effluent gas scrubbing or to transport residual solids should be recycled and/or treated to produce an essentially pollution-free effluent. New incinerators in the U.S. are now normally built to recover heat in the form of steam, instead of discharging the combustion heat to the atmosphere as hot flue gas.

Refractory Incinerators

Most of the incinerators which have been built in the U.S. do not practice energy recovery. They utilize a refractory furnace where the solid waste is burned with air. The furnace may be a fixed hearth type or an inclined rotary kiln. Excessive temperatures are avoided by using a quantity of air in excess of that theoretically required for combustion, the excess air serving as a cooling medium. Average furnace exit temperatures are usually in the range 760-1010°C (1400-1850°F).

Grates are provided in fixed hearth furnaces as a passage for underfire air, while supporting the solid waste being burned. The most common of the designs available are the many types of moving grates which transport the solid waste and residue through the furnace and, at the same time, promote combustion by inducing agitation and passage of underfire air.

Incinerators with Heat Recovery

The simplest form of energy recovery is the use of a waste heat boiler with a conventional refractory incinerator, that is extracting heat from the flue gases usually to make low pressure steam. A more effective type of heat recovery unit utilizes furnace walls made of closely-spaced steel tubes welded together, with water or steam circulated through the tubes to extract heat generated during combustion. This procedure not only leads to heat recovery, but allows a major reduction in air requirements, thus reducing the size of air pollution control equipment and other facilities. Where high pressure steam is made, it can be used to drive turbines for electric power production. The decision as to energy recovery is governed primarily by the nature of the market for steam, including demand patterns and potential value.

Slagging Incinerators

If the combustion air flow for a given burning rate of solid waste in a refractory incinerator is reduced, the combustion temperature increases.

At a combustion temperature of about 1600°C (2912°F), a molten residue is obtained which, when cooled, provides a dense inert material useful as a landfill. The other advantages of this reduction in combustion air flow are the reduced gas volume, simplifying air pollution control, and the more effective combustion at high temperature. Slagging, high temperature incinerator systems are in the development stage.

Suspension-Fired Incinerators

Extensive size reduction of the solid waste allows furnace designs analogous to pulverized coal steam boilers commonly used by the electric utilities and large industrial plants. The pulverized waste is suspended in an air stream and introduced into the combustion zone, where burning is very rapid. Unlike more conventional incinerators, the major portion of the residue or flyash is carried by the hot flue gases out of the combustion zone directly to the air pollution control solids recovery equipment. A combination refuse/coal combustion system which takes advantage of existing coal-fired suspension type boilers has been undergoing large-scale demonstration.

Fluidized Bed Incinerators

The fluidized bed is a special form of suspension-fired incinerator where the combustion is carried out in the presence of a suspended bed of inert solids whose behavior is analogous to that of a fluid. The fluidized bed aids contact between the air and the solid waste, improving combustion. Agglomeration of the flyash may also be promoted, improving particulate recovery. A system for extracting electric power by expansion of flue gases from an elevated pressure fluidized bed is under development.

PYROLYSIS

As discussed earlier in this chapter, little or no air is introduced into the elevated temperature pyrolysis chamber. Instead of combustion, a complex series of decomposition and other chemical reactions take place. Pyrolysis of municipal solid waste produces low sulfur gaseous, liquid and solid products which are potentially useful as fuels or chemical raw materials. The nature of these products depends primarily on the composition of the waste, pyrolysis temperature, pressure and residence time.

Also, unlike incineration which is highly exothermic, the addition of heat to the pyrolysis chamber is usually necessary. The method of heat introduction is a major distinguishing factor between various pyrolysis processes. For example, auxiliary fuel combustion, highly preheated air, circulating heated solids, and limited oxygen introduction to produce heat by oxidizing part of the carbon present in the waste have all been used.

The elimination of inorganic constituents of the solid waste is a useful step in pyrolysis processes to avoid contamination of products. Therefore, separation steps to recover glass and metal by-products fit naturally into pyrolysis schemes.

The control of air pollution in pyrolysis processes is eased to some extent, as compared to incineration, because of the reduced volume of gases to be treated. However, water pollution control problems may be more serious than in incineration due to extensive production of water and water-soluble organic chemicals which must be disposed of. This is a particularly difficult problem in low temperature pyrolysis where liquid yields are high.

REFERENCES

1. DeMarco, J. *et al.* Municipal-Scale Incineration Design and Operation. PHS Publication No. 2012, Washington, D.C. U.S. Government Printing Office. 1969 (formerly Incinerator Guidelines-1969).
2. Franklin, W. E. *et al.* Resource and Environmental Profile Analysis of Solid Waste Disposal and Resource Recovery Options. Midwest Research Institute. Kansas City, Mo. 1974. 28 pages.

CHAPTER 2

RECOVERY AND UTILIZATION OF ENERGY

While common in Europe, the conversion of solid waste to energy in the U.S. was until recently only an interesting idea reduced to practice in a very few plants.[1] However, the recent awareness of an "energy crisis" has spurred no less than 20 cities to consider projects for steam generation by prepared refuse combustion, many using existing fossil fuel-fired steam boilers. The feasibility of this approach has been shown in a project partially supported by the U.S. Environmental Protection Agency.[2] Most of the following discussion deals with incinerators specially designed for steam production. The use of prepared refuse to supplement coal or other fuels is dealt with specifically in a later section of this chapter.

ENERGY RECOVERY SYSTEMS *VS* REFRACTORY INCINERATORS

Non-energy recovery incinerators have refractory combustion chambers, while combustion chambers in steam-producing incinerators are usually water tube wall construction. The choice between burning refuse in refractory incinerators or providing for energy recovery is generally not clear cut. Advantages and disadvantages of both approaches are compared in Table 2.

Pyrolysis provides another alternative for energy recovery in the form of fuels. Pyrolysis processes which produce liquid fuels are particularly attractive, from the point of view of energy recovery, because of the ease with which this form of energy can be stored.

Generally, energy recovery systems cost more to install and operate, present more operating difficulties and safety considerations, and put the municipality into a business, but do provide significant credits from the sale of the steam. It should be noted that steam is not a storable

Table 2. Comparison of Refractory Incinerators and Energy Recovery Systems

Refractory Incinerators	Energy Recovery Systems
1. High excess air required to control furnace temperature	1. Moderate excess air
2. Large refractory combustion chamber to handle high gas flow	2. Moderate size combustion chamber, but requires water tube wall construction, and may require additional parallel lines for steam supply reliability
3. Costly furnace auxiliary equipment due to high gas flow, including FD and ID fans, air pollution control equipment, ducts, and stack	3. Furnace auxiliary equipment similar to refractory incinerators, but less costly due to lower gas flow
4. No steam facilities required	4. Requires expensive steam facilities and controls, including water tube walls, waste heat boiler, boiler feedwater treating, soot blowers, steam condensers, and steam distribution
5. Requires flue gas cooling system such as spray chamber (waste heat boiler sometimes used)	5. Waste heat boiler producing steam used for cooling
6. Relatively simple operating procedures	6. Operations complicated by necessity to meet steam supply demands, maintenance of steam equipment, presence of high pressure steam systems, etc.
7. Moderate potential for corrosion, especially in air pollution control equipment	7. Possible steam tube corrosion and erosion require monitoring and additional maintenance cost; air pollution control equipment corrosion can be problem as with refractory incinerators
8. Moderate operating costs	8. Higher operating costs because of increased complexity
9. Only possible by-product credits are for pre- or post-incineration salvage	9. Considerable steam credits possible in addition to salvage, including in-plant use of steam for fan drives, heating, etc.

commodity, and major customers should be contracted before a project is justified by the steam credits. Major investments are required for generation and distribution systems.

To illustrate the potential for energy recovery, steam production in several recently designed installations is shown in Table 3. For a given incinerator, the amount of steam generated is primarily a function of the combustible content of the refuse, as shown in Table 4.

Table 3. Examples of Reported Steam Generation Quantities[3]

	System 1	System 2	System 3	System 4
Solid Waste Type	A	B	A	A
Steam Temperature, °C	327	205	260	241
(°F)	(620)	(401)	(500)	(465)
Steam Pressure, atm abs.	28.2	18.0	16.3	18.7
(psig)	(400)	(250)	(225)	(260)
Steam Production, tons/ton refuse[a]	3.6	4.2	1.4-3.0	1.5-4.3[b]

[a]Tons steam per ton of refuse fed to the furnace varies with refuse heating value, as shown in Table 4.

[b]See Table 4 for variation of steam production with heating value.

A. Solid waste as received.

B. Refuse prepared by shredding and partial metal removal.

Table 4. Effect of Solid Waste Heating Value on Steam Production[4]

Nominal Refuse Heating Value, cal/g (BTU/lb)	3611 (6500)	3333 (6000)	2778 (5000)	2222 (4000)	1667 (3000)
Refuse, % Moisture	15	18	25	32	39
% Noncombustible	14	16	20	24	28
% Combustible	71	66	55	44	33
	100	100	100	100	100
Steam Generated, tons/ton refuse[a]	4.3	3.9	3.2	2.3	1.5

[a]These values are calculated for a specific operating incinerator, but they are believed to truly reflect actual results.

ENERGY USES

It should be noted that the primary purpose of a thermal processing facility is to dispose of municipal solid waste. This waste never stops coming. Therefore, any complications which tend to reduce reliability must be carefully evaluated. Both in-plant use and export of steam should be considered, as well as the use of steam turbines to generate electric power, although the latter is not now a significant method for energy recovery in the U.S. As will be described, air conditioning can also be exported.

In-Plant Uses

The most obvious and efficient use of the energy recovered is in the thermal processing plant itself. Power plants, oil refineries, and chemical and other large manufacturing facilities have long practiced "energy recycle" in order to keep the overall utility (*i.e.*, fuel, water, electricity) costs to a minimum.

Steam turbines to drive induced and forced draft fans, large pumps, and other significant power requirements provide a major outlet for recovered energy. Other uses include space heating, snow and ice melting from ramps, and tracing to prevent freezing of water lines. One steam generating incinerator estimates its steam usage as tabulated in Table 5.

Table 5. Example of In-Plant Usage of Steam[a]

	% of Total Steam Generated
Fans	25.3
Feedwater Pumps	5.1
Feedwater Heater	2.1
Space Heating	5.3
Water Heating	2.7
Condenser Protection against Freezing	6.1
Total In-Plant Usage	46.6
Maximum Available for Sale	53.4

[a]Calculated from Reference 5.

All the energy requirements of the plant are not usually supplied by the steam. Steam-driven equipment has a higher capital cost; therefore the steam is not "free." Backup electric-driven equipment must be provided to start-up fans and other equipment when steam is not available. Weather protection, heating, and other services must be provided, even when steam-generating facilities are not operating. Obviously, the necessary backup equipment adds to capital costs. Figure 1 depicts an example of an in-plant steam distribution scheme.

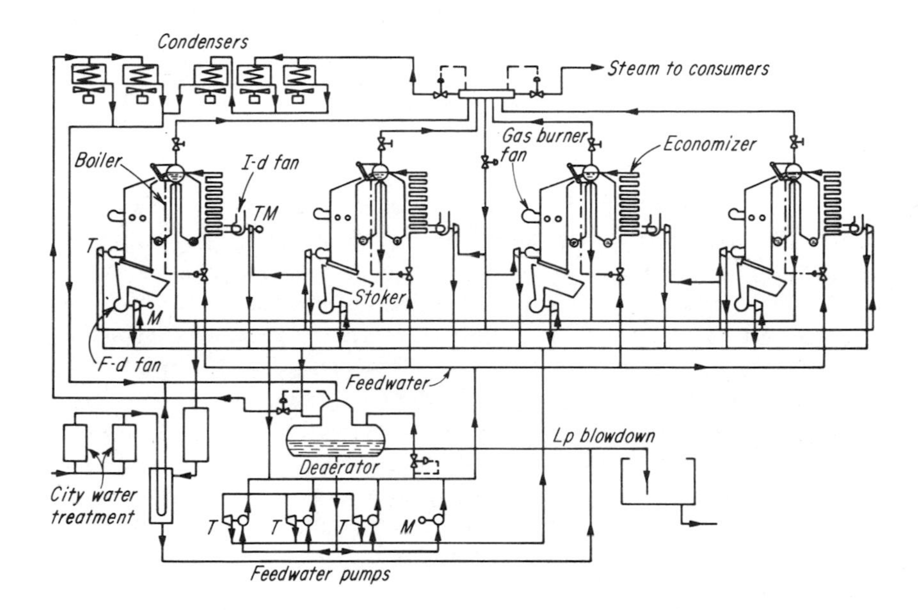

Figure 1. In-plant steam distribution scheme. Auxiliary equipment of the four combinations of incinerators and boilers in Chicago is steam-turbine driven, except for the electric motors needed to start the plant when no steam is available. Note use of air-cooled condensers.[5]

T = steam turbine; M = electric motor

In-plant usage of steam is a straightforward method of utilizing up to about half of the energy available from combustion of the solid waste. No capital cost is required for external steam distribution, and the complexity of matching steam supply to export demand is avoided.

Steam Export

Because of projected energy shortages and ever-rising fuel costs, the economic picture has begun to favor generation and sale of energy produced from solid waste. This is illustrated by the simple "fuel equivalent" comparison made in Table 6.

Table 6. Comparison of Relative Values of Refuse and Fuel Oil Based on Heats of Combustion

Crude Oil Price, $/barrel	3.50	7.00	10.00	15.00
Equivalent "Value" of Refuse, $/metric ton[a]	6.80	13.61	19.44	29.16

[a]Based on 42 gallons of oil per barrel, 0.9 specific gravity, 10,000 calories per gram (18,000 BTU/lb); and refuse at 2778 calories per gram (5000 BTU/lb).

However, it cannot be emphasized enough that these "favorable economics" can be illusory. No credits will accrue if the energy cannot be sold. Much, if not most, of the steam produced in U.S. solid waste incinerators is being condensed because of lack of customers.

Steam generated can be used directly for heating purposes. Typically, the steam pressure as generated is in the 18-45 atm (250-650 psig) range. If low pressure steam, for example 11 atm (150 psig), can be sold for space heating, the pressure difference between the high and low pressure steam can be utilized to drive in-plant noncondensing turbines prior to entering the distribution systems. The use of steam for heating may require both steam distribution and condensate return lines. If the consumer is distant, the condensate return line may be eliminated, but this increases the cost of boiler feedwater treating.

Obviously this type of arrangement creates a responsibility for the municipality to reliably deliver the steam, the failure of which may have drastic consequences. Therefore, the use of auxiliary burners fired by fossil fuel and/or an auxiliary fossil fuel-fired package boiler is required to meet demands during downtime of the thermal processing facility, during periods of wet refuse, and sometimes for peak loads. An auxiliary burner can provide continuity of steam supply during periods of refuse feed equipment failure or other up-stream problems. The standby package boiler is used when the thermal processing facility is totally inoperable. An increased number of parallel furnace lines may also be required to improve steam supply reliability.

The use of waste lubricating oils and other waste oils have been proposed as auxiliary fuels for steam-generating incinerators.[6] However, such use requires precautions against hazardous contamination, such as flammable solvents; consideration of the possibility of steam tube fouling due to metallic impurities; and provisions for controlling particulate emissions due to metallic impurities, which may include lead contents on the order of 1% in raw waste oils.[7]

Ideally, of course, the consumer has an alternative steam supply available, and therefore is not dependent on the solid waste-generated steam. In this case, the steam will usually be less valuable to the consumer and lower prices will be obtained.

Air Conditioning

One recent project uses steam to heat downtown office buildings in the winter and to cool them in the summer.[8,9] This idea of "District Heating and Cooling" is not new, having been practiced elsewhere, but the use of solid waste as the primary fuel is novel in the U.S. The additional responsibility for supplying cooling in the form of chilled water further complicates the disposal task. Figure 2 depicts a complete system for accomplishing disposal, heating and cooling.

In this plant, pumps are driven by noncondensing steam turbines, and water chillers are driven by condensing turbines using steam exhausted from the pump drives. The system is capable of producing 11.2 atm absolute (150 psig) saturated steam for heating and 5°C (41°F) chilled water for cooling. A condenser is available to handle surplus steam; and standby package boilers are provided to supplement steam demands, in order to increase overall plant reliability and operating flexibility during maintenance periods. Further detail is provided in Table 7.

Design and operation are complicated by peak daily demands and peak demands in winter for heating and in summer for chilling. Even for production of steam for both heat and chilling, the sum of the peak demand in the spring and the fall is less than individual demands for winter or summer. Since an incinerator normally must handle a specific daily tonnage of waste and will produce a fixed steam rate regardless of demand, it is evident that very careful analysis of supply and demand during the project planning phases is imperative.

Other Energy Uses

Marketing steam and chilled water to municipally-owned buildings and university complexes for heating and cooling may be easiest institutionally, but industrial markets for process steam where demand is uniform, without exaggerated peaks and valleys, is much more satisfactory from the point of view of incinerator operation. Possible industrial consumers include power plants, oil refineries, chemical plants, and other plants utilizing low-level energy (*i.e.*, steam rather than direct firing of fuel, such as in glass or steelmaking furnaces).

Other process possibilities include municipal sewage sludge drying. Due to increasing environmental restrictions on ocean dumping and landfill

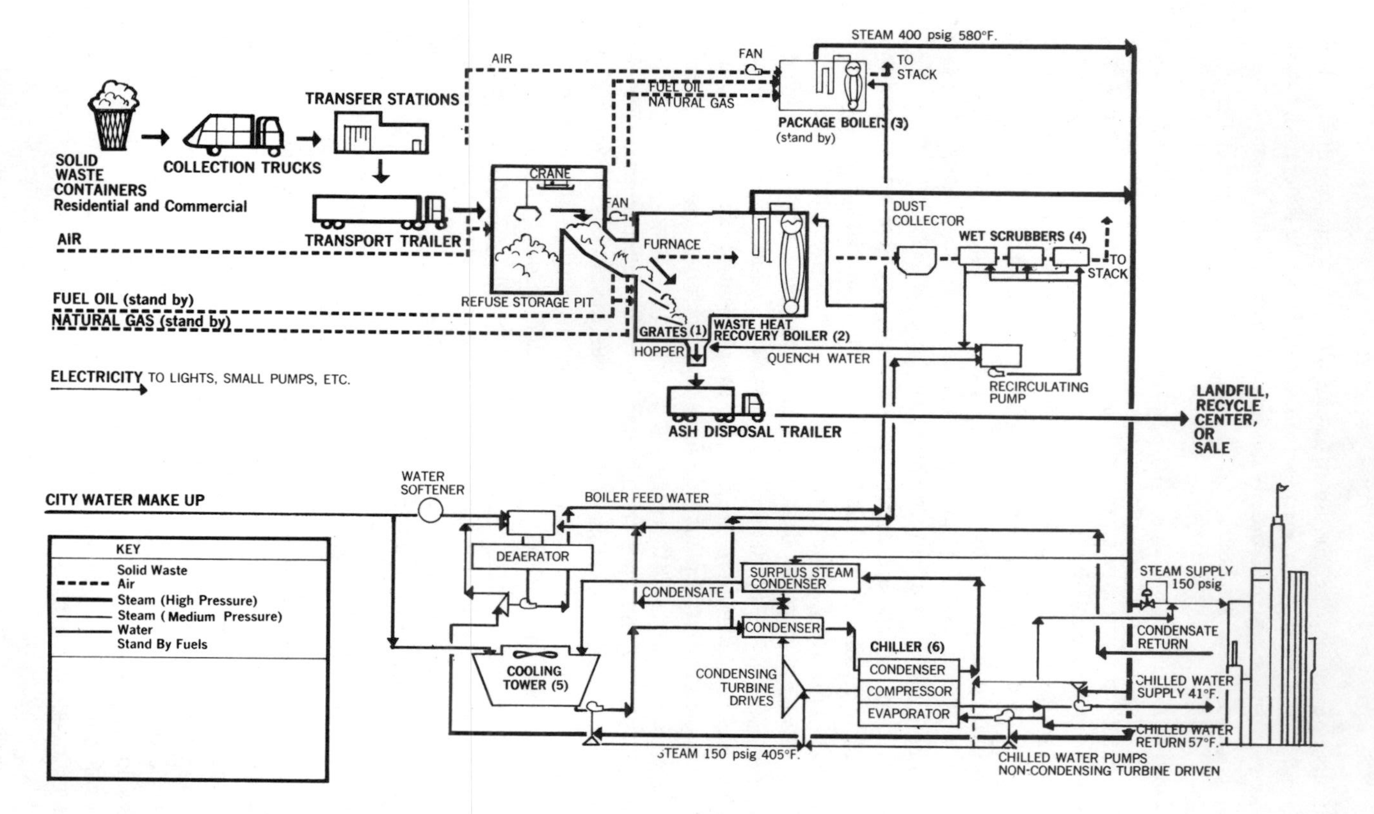

Figure 2. Simplified flow diagram—central heating and cooling plant fueled by solid waste.[9]

Table 7. Pertinent Data for Heating and Cooling District Supplied by Solid Waste-Generated Steam[8,9]

Total City Refuse[a]	52.9 MT/hr	(1400 ST/day)
Number of Incinerators	2	
Capacity per Incinerator	13.6 MT/hr	(360 ST/day)
Number of Buildings Served	40	
Steam Generation Capacity (Incinerators)	97.5 MT/hr	(215,000 lb/hr)
Standby Steam Boiler Capacity	56.7 MT/hr	(125,000 lb/hr)
Steam Pressure as Generated	28.2 atm abs	(400 psig)
Steam Temperature as Generated	316°C	(600°F)
Steam Pressure (Saturated) Supplied for Heating and Condensing Turbines	11.2 atm abs	(150 psig)
Chilled Water Capacity	40.8×10^6 Kcal/hr	(13,500 tons of refrigeration)
Chilled Water Supply Temperature	5°C	(41°F)
Chilled Water Return Temperature	14°C	(57°F)
Length of Distribution Pipeline	4.57 kilometers	(15,000 ft)

[a]Parts of the distribution system and certain main plant components are designed for the projected ultimate plant capacity of 56.7 MT/hr (1500 ST/day) using five incinerators.

operations, sewage sludge incineration is growing in importance. Use of hot flue gases or solid waste-generated steam to predry the sludge results in economic and energy savings. Co-incineration of the dried sludge, or even the wet sludge, is also an interesting approach which has been used in the past to a limited extent.

Fresh water production from seawater or brackish water by distillation or other desalting processes requires energy which can be supplied from thermal processing facilities. This use, which can be considered in arid areas or other areas with special water problems, *e.g.*, Southern California and Long Island, may be attractive because the pure water can be stored, allowing steady use of energy produced by the thermal processing facility.

Electricity can be produced from solid waste in three ways:

1. Generating steam to drive electrical generators.
2. Generating steam to sell to power companies.
3. Mixing prepared refuse with fossil fuel in power plant boilers.
4. High pressure, high temperature incineration using exhaust gases to drive an expansion turbine-generator.

While several European incinerators use steam produced to generate electricity, this approach is rare in the U.S. The problems of inefficient generation from the relatively low pressure steam normally produced and the difficulties and cost in reliably producing high pressure, high temperature steam in solid waste incinerators appear to be formidible deterrents to this approach, though future improvements may be expected. Similarly, the sale of low pressure steam to electrical power companies is not usually attractive because steam pressures near 136 atmospheres (2000 psi) are preferred for efficient generation of electricity. Combined firing of prepared refuse with fossil fuel, which is attractive, will be covered in a later section of this chapter. Direct production of electricity by incineration of prepared solid waste in a pressurized fluidized bed, and expansion of the hot gases through a turbine, though promising, has not yet reached commercialization.[10]

ENERGY RECOVERY SYSTEMS

Energy recovery may be accomplished in such diverse systems as pyrolysis processes, high pressure fluidized beds with gas turbines, and other methods which have not been demonstrated on a commercial scale. The systems described here generate steam by combustion of solid waste.

At least four general types of steam generation systems can be employed:

- Grate furnace with refractory walls and a waste heat boiler
- Grate furnace with water tube walls and a waste heat boiler
- Suspension-fired steam boiler
- Combined firing of prepared refuse with fossil fuels

In the first two types, it is not necessary to shred or otherwise prepare the solid waste. In the latter two, preparation is required.

Refractory Incinerator with Waste Heat Boiler

This is a conventional incinerator with refractory walls followed by boiler tubes erected in the flue gas stream leaving the furnace. While this approach obviously produces energy credits, the full benefit of energy recovery is not achieved because considerable excess air must be used to control temperature in the combustion chamber. This excess air reduces the efficiency of energy recovery by carrying heat from the furnace system up the stack, and also requires larger downstream equipment such as fans and air pollution control equipment based on the high flue gas rate emitted. Careful design of boiler tubes is necessary to control temperature and to avoid slagging and corrosion.

Incinerator with Water Tube Walls

In the water tube wall incinerator firing unprepared waste, shown in Figure 3, heat is recovered directly from the combustion zone, eliminating the need for large quantities of excess air used for cooling in refractory incinerators. This is accomplished by the use of water tube walls in place of refractory, where water is circulated to remove heat from the combustion zone. This approach not only recovers additional energy, but reduces flue gas quantities significantly, reducing the size of downstream equipment. For example, the use of water tube walls may be expected to reduce flue gas quantity by 30-40% over a refractory furnace operating at the same temperature. Excess air requirements are usually on the order of 40-80% as compared to 100-200% often encountered in refractory furnaces.

Due to advantages of energy recovery and flue gas volume reduction, the newest steam-generating incinerators have been of the water wall type. Tables 8-11 provide actual operating data for such an incinerator.

Suspension-Fired Steam Boiler

The suspension-fired steam boiler, shown in Figure 4, is based on the design of pulverized coal boilers commonly used throughout the world for electric power production and industrial boilers. Suspension firing has been used for other waste materials such as bark and bagasse,[13] but only recently has shredded waste been burned in suspension boilers.[14-16] In one, about 50% of the combustion is accomplished in suspension, with the remaining combustion occurring on moving grates.[15,16] In another, the prepared waste is burned with coal,[14] as described in the next section. A third suspension-fired boiler is operated within an industrial plant.

Prepared Refuse Combustion in Existing Boilers

A project which has created great interest on a prototype scale (125-megawatt boiler) is the firing of prepared refuse in mixture with pulverized coal in an existing boiler. A commercial scale facility is planned by the participants. The attractive features of this approach include:

1. The boiler units are already in place, obviating the need for new thermal processing facilities.
2. Air pollution from sulfur oxides is reduced because the prepared refuse is normally lower in sulfur than the coal it replaced. Overall particulate emissions may also be reduced, compared to separate incineration, because of the separation of inorganic materials from the combustibles, and efficient electrostatic precipitators on the boiler.

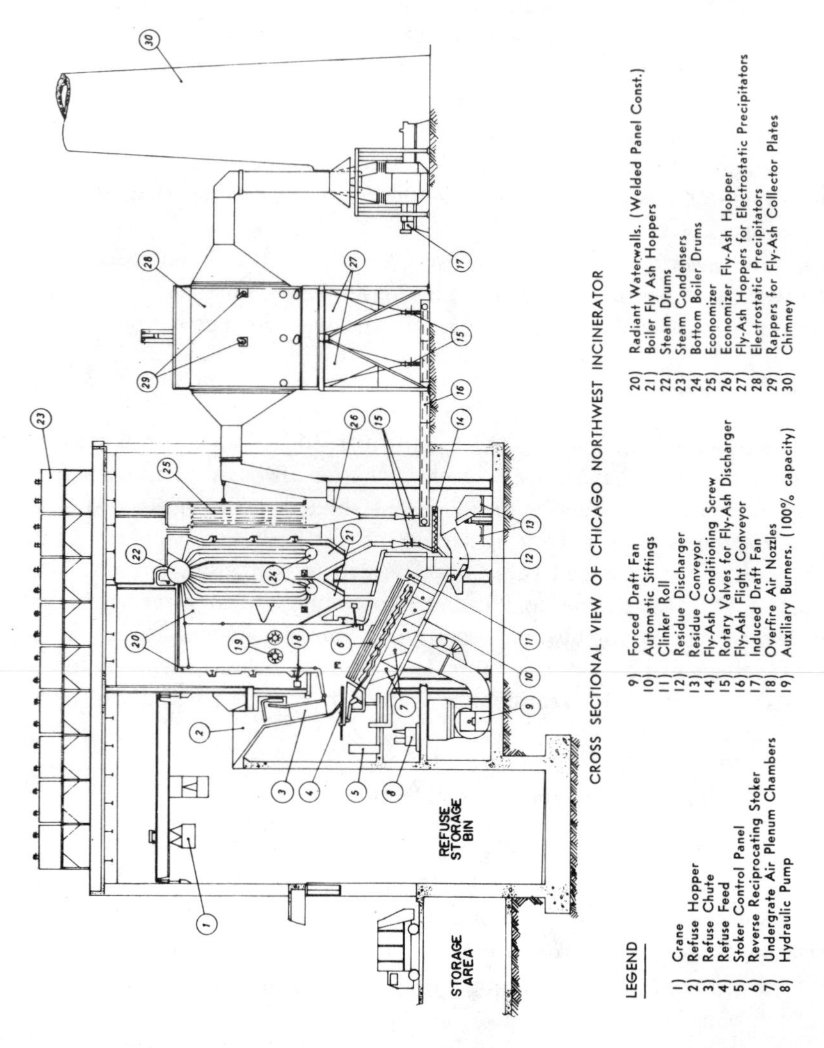

Figure 3. Chicago Northwest Steam Generating Incinerator.[11]

Table 8. Quantities and Flue Gas Analysis for Steam-Generating Incinerator[11]

Heating Value of Refuse	2422 cal/g	(4360 BTU/lb)
Incinerator Capacity	15.2 MT/hr	(401 ST/day)
Refuse Firing Rate	15,166 kg/hr	(33,434 lb/hr)
Heat Input	36.7×10^6 Kcal/hr	(145,772,000 BTU/hr)
Gas Exit Temperature	211°C	(411°F)
Ambient Air Temperature	22.8°C	(73°F)
Gas Composition, Vol %		
CO_2	10.49%	
O_2	9.02%	
CO	0.0%	
N_2	80.49%	
Excess Air	71.7%	

Table 9. Boiler Losses and Efficiency for Steam-Generating Incinerator[11]

Heating Value of Refuse	2422 cal/g	(4360 BTU/lb)
Heat Losses		
Dry flue gas		11.40%
Moisture in fuel		4.01
Moisture in air		1.22
Moisture from burning hydrogen (H_2)		8.83
Combustible in residue		2.83
Moisture in residue		0.30
Moisture flashed from quench		0.32
Radiation loss		0.41
Unaccounted for losses		1.50
Total losses		30.82%
Efficiency		69.18%

Table 10. Air and Gas Quantities for Steam-Generating Incinerator[11]

Heating Value of Refuse	2422 cal/g	(4360 BTU/lb)
Refuse Firing Rate	15,166 kg/hr	(33,434 lb/hr)
Dry Gas per Weight of Refuse	6.14 kg/kg	(6.14 lb/lb)
Total Moisture per Weight of Refuse	0.516 kg/kg	(0.516 lb/lb)
Weight of Gas per Weight of Refuse	6.656 kg/kg	(6.656 lb/lb)
Gas Temperature	211°C	(411°F)
Density of Gas	0.734 kg/CM	(0.0458 lb/CF)
Density of Water Vapor	0.457 kg/CM	(0.0285 lb/CF)
Total Weight of Dry Gas	93,117 kg/hr	(205,285 lb/hr)
Total Weight of Water Vapor	7825 kg/hr	(17,252 lb/hr)
Total Weight of Products of Combustion	100,942 kg/hr	(222,537 lb/hr)
Volume of Dry Gas at 211°C (411°F)	2116 CM/min	(74,703 CF/min)
Volume of Water Vapor at 211°C (411°F)	282 CM/min	(9,950 CF/min)
Volume of Products of Combustion at 211°C (411°F)	2293 CM/min	(80,980 CF/min)
Available I.D. Fan Capacity at 260°C (500°F)	4030 CM/min	(142,300 CF/min)
Weight of Air for Combustion	85,192 kg/hr	(187,815 lb/hr)
Air Volume at 21.1°C (70°F)	1182 CM/min	(41,750 CF/min)
Available F.D. Fan Capacity at 21.1°C (70°F)	2379 CM/min	(84,000 CF/min)

Table 11. Overall Energy Recovery Performance of Steam-Generating Incinerator[11]

Heating Value of Refuse	2422 cal/g (4360 BTU/lb)
Steam Generation per Weight of Refuse	2.98 kg/kg
Overall Efficiency of Incinerator and Boiler	69.18%
Heat Loss Due to Combustibles in Residue	2.83%
Stack Gas Temperature	211°C (411°F)

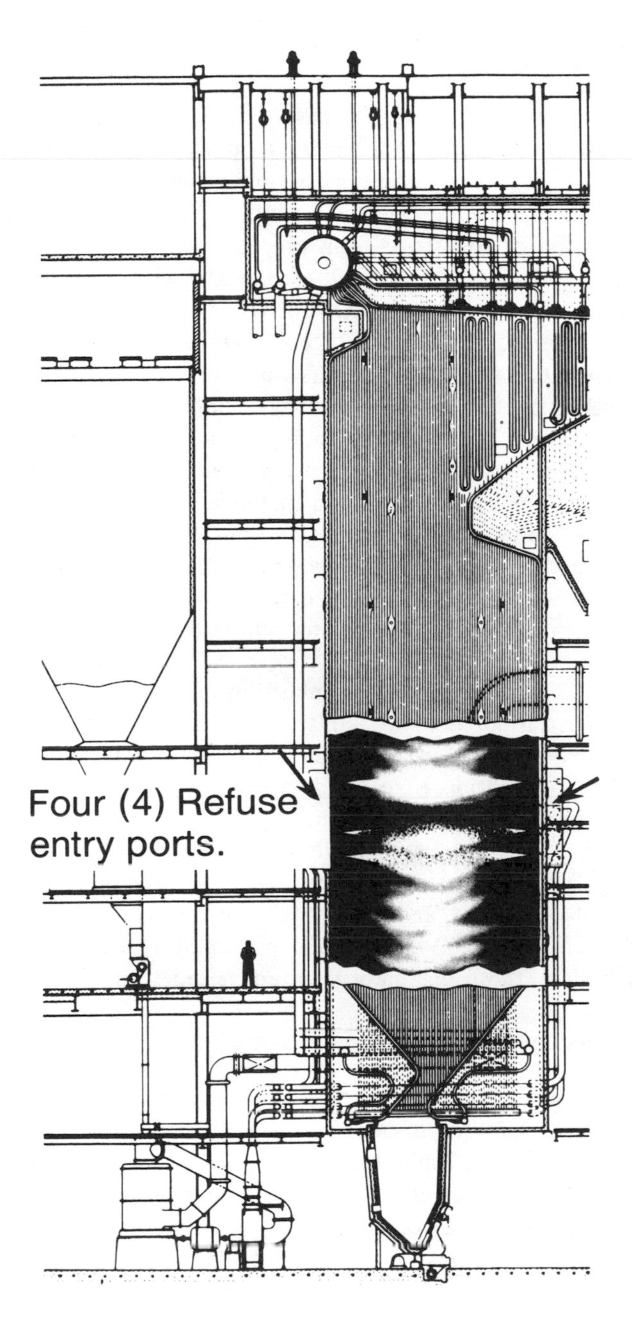

Figure 4. Suspension-fired boiler.[12]

3. Fossil fuel consumption for power generation will be reduced.
4. Facilities and markets already exist for the electrical energy produced.

The technique is based on the premise that if solid waste is prepared so that its flow characteristics are similar to pulverized coal, and its ratio to coal-fired is kept low enough, the coal boiler will not be significantly affected and the above advantages will accrue. Tests to date show that efficient generation of electricity from prepared refuse can be accomplished by this approach. A single 500-megawatt boiler operated at 75% use factor could help dispose of 340,000 ST of solid waste per year, saving on the order of 150,000 ST of coal per year.

From the results obtained, it appears that on the order of 10-20% of the heating value fired in a normal coal-fired boiler can be supplied by solid waste which has been shredded and processed to recover noncombustibles such as metals and glass. These processing steps are expensive, but partially paid for by the recovered materials.

Particulate emission tests for combined firing have been inconclusive. The data obtained suggest possible problems with electrostatic precipitator performance when burning the prepared refuse with low sulfur coal. Such problems are common when burning low sulfur coal alone, but are costly to resolve.

The combined fossil fuel/refuse combustion approach requires additional demonstration in various types of boilers over an extended period of time before it can become a fully acceptable outlet for municipal solid waste disposal. Close observation is required to ensure that no unusual corrosion or other detrimental effects occur.

The possibility of firing prepared refuse in oil-fired boilers is now undergoing study. This presents special problems of refuse feeding techniques, disposal of residue, and particulate emission control, since adequate solids handling facilities and electrostatic precipitators are available on only a few boilers, those that have been converted from coal to oil firing.

Refuse-to-Energy Projects

Thirty-three projects in operation or in the planning stage are listed in Table 12.[15] Details on refuse-fired systems operating or under construction are provided in Table 13.[15]

A recent U.S. Environmental Protection Agency report provides an up-to-date list of municipalities committed to or having expressed an interest in resource recovery systems, most of which involve energy recovery.[17]

Table 12. Refuse-to-Energy Projects Operating or Planned[a]

Generate Steam[b] for Off-Site Use: Location (Owner/Operator)	Status
Montreal, Que., Canada (Muni)[c]	Op
Quebec, Que., Canada (Muni)	Op
Hamilton, Ont., Canada (Muni)[c]	Op
Harrisburg, PA (Muni)[c]	Op
Chicago, IL (Muni)[c]	Op
Nashville, TN (NPC)	Op
Braintree, MA (Muni)	Op
Norfolk, VA (U.S. Navy)	Op
Saugus, MA (PriC)	UCon
Portsmouth, VA (U.S. Navy)	UCon
Akron, OH (Muni)	DSC
Albany, NY[d]	DSC
Cleveland, OH (Muni)	FSC
Palmer Twp, PA (Muni)	FSC
Generate Steam[e] to produce electricity: Location[f] (Utility)[g]	
St. Louis, MO (Union Electric)	Op[h]
Brockton, MA[i]	SS
Chicago, IL (Commonwealth Edison)	UCon
Ames, IA (Muni)	SS
Bridgeport, CT (Northeast Utilities)	CAwd
Hempstead, NY (Muni)[j]	CAwd
New Britain, CT (Muni)	CNeg
Monroe Cnty, NY (Rochester G&E)	CNeg
Lane Cnty, OR (Muni)	PDC
Memphis, TN (TVA)	PD
Hackensack Meadowlands, NJ (Public Service E&G)	FSC
Milwaukee, WI (Wisconsin Electric)	FSC
Wilmington, DE (Delmarva P&L)	US
Washington, DC (Pepco)	US
Montgomery Cnty, MD (Pepco)	US
Madison, WI (Madison G&E)	US
Los Angeles, CA (Muni)	US
Honolulu, HI (Hawaiian Electric)	US
Housatonic Valley, CT (Muni)	US

CAwd – Contract awarded
CNeg – Contract being negotiated
DSC – Design study complete
FSC – Feasibility study complete
Muni – Municipal government
NPC – Nonprofit corporation
Op – Operational
PF – Preliminary design underway
PDC – Preliminary design complete
PRiC – Private corporation
SS – System shakedown in progress
UCon – Under construction
US – Under study

[a]As reported by the U.S. Environmental Protection Agency, October 1974.
[b]From processed or raw refuse in a waterwall boiler or waterwall incinerator.
[c]Steam available for sale, but no contract signed yet.
[d]City to own refuse processing plant which contractor will operate; state to own and operate steam production facility.
[e]From processed waste–that is, shredded refuse with heavy fraction (metals, glass, etc.) removed or prepared refuse fuels in powdered, pellet or briquette form.
[f]Location of refuse processing plant.
[g]In some cases, the utility may not have signed a contract to burn the prepared refuse in its boilers, but it is participating in engineering and testing aspects of the project.
[h]Demonstration plant operational, expansion planned.
[i]Preparation facility here already has a major paper company under contract for about 70,000 tons of its fuel product over a 10-yr period.
[j]Sale of electricity to Long Island Lighting Co.

Table 13. Principal Equipment in Refuse-Fired Power Plants Operating or Under Construction in the United States and Canada

Solid Waste Recovery System, Ames, IA	
Owner	City of Ames
Consulting engineer	Gibbs, Hill, Durham & Richardson Inc.
Expected startup	June 1975
Primary-shredder feed conveyor - 7 ft wide	Mayfran Inc.
Shredders, 2	American Pulverizer Co.
One primary, one secondary; each 50-tph capacity, 5-ft-diam wheel X 7.5-ft rotor; 1000-hp, 4160-V, 720-rpm motor drive	
Shredder discharge conveyors, 2	Rexnord, Carrier Div.
One per shredder; each oscillating type, 6 ft wide, 15-hp motor drive	
Second-stage feed conveyor (standby)	Rexnord, Carrier Div.
Vibrating type, 4 ft wide, 7.5-hp motor drive. Normally, an inclined belt conveyor, 5 ft wide, would be used to convey refuse to the second shredder.	
Magnetic-metal separators, 3	Dings Co.
One 5 ft wide, two 2-ft-wide magnetic belt pulleys	
Air classification system	Rader Pneumatics Inc.
Includes one 600-cu ft surge bin complete with scalping roll and flight conveyor; vibrating feeder, 8 ft wide X 10 ft long; refuse conveying pipe, 42 in. diam; cyclone separator, 14 ft diam X 36.5 ft high; exhaust fan with 200-hp motor drive, etc.	
Pneumatic conveying systems, 5	Pneumatic Systems Inc.
Process plant to storage: One system consisting of a positive-displacement blower rated 6545 cfm at 4.53 psi, 200-hp motor drive; inlet silencer; piping; feeder; cyclone separator, etc. Storage to boilers; Four identical and separate systems each consisting of a rotary positive-displacement blower rated 2390 cfm at 3.51 psi, 60-hp motor drive; silencer; feeder; inlet feed chute; fluffing roll, etc.	
Refuse storage bin–500-ton capacity, 84 ft diam	Atlas Systems Corp.
Noncombustibles separation system	Combustion Power Co.
Separation-system conveyors, 12–Various sizes	Fairfield Engineering Co.
Separation-system bucket elevators, 5	Fairfield Engineering Co.
Boilers (existing), 3	Combustion Engineering Inc., Riley Stoker Corp., Union Iron Works Co.
One 360,000 lb/hr at 900 psig/900 F with tangential burners and electrostatic precipitator; one 125,000 lb/hr at 725 psig/825 F with traveling-grate spreader stoker and multiple-cyclone dust collector; one 95,000 lb/hr at 720 psig/825 F with traveling-grate spreader stoker and multiple-cyclone dust collector	
Solid Waste Utilization System, St. Louis, MO	
Owner	City of St. Louis/Union Electric Co.
Consulting engineer	Horner & Shifrin Inc.
Commercial operation	April 1972
Equipment at City of St. Louis processing facility:	
Raw-refuse receiving conveyor–84 in. wide	Te-Co Inc.

Table 13, continued

Belt conveyors	Continental Conveyor Co.
Vibrating feeders and conveyors	Borg-Warner Corp., Stephens-Adamson Div.
Shredder	Gruendler Crusher & Pulverizer Co.
5-ft-diam wheel X 6.7-ft rotor, 1250-hp motor drive	
Air classification system–45 tph capacity	Rader Pneumatics Inc.
Pneumatic transport equipment	Rader Pneumatics Inc.
Storage bin–300-ton capacity	Miller Hofft Inc.
Nuggetizer (hammermill for densifying ferrous scrap)	Eidal Corp.
Equipment at Union Electric Co.'s Meramec Station:	
Receiving bin–100-cu yd capacity	Miller Hofft Inc.
Pneumatic transport equipment	Rader Pneumatics Inc.
Storage bin	Atlas Systems Corp.
Boilers (existing), 2	Combustion Engineering Inc.
Each 925,000 lb/hr, single-reheat type with tilting tangential burners (at four elevations in each corner of the furnace) and electrostatic precipitators. The furnace is about 28 ft X 38 ft in cross section, with a total inside height of about 100 ft.	
Solid Waste Reduction Unit, Hamilton, Ont., Canada	
Owner	City of Hamilton
Consulting engineer	Gordon L. Sutin & Assoc. Ltd.
Commercial operation	Summer 1972
Incineration capacity	600 tpd
Refuse-pit size	40 ft X 80 ft X 30 ft deep
Refuse-pit apron conveyors, 4	Rex Chainbelt Ltd (Can)
Pulverizers, 4–Each vertical-shaft-type, 200-hp motor drive	Heil Co.
Belt conveyors	Rex Chainbelt (Canada) Ltd
Magnetic-metal separators, 2	Eriez Magnetics, Dings Co.
Shredded-refuse storage bin–70 ft diam.	Atlas Systems Corp.
Shredded-refuse fuel supply systems, 2	Detroit Stoker Co.
Each (one per boiler) consists of a swinging distribution spout and three parallel pneumatic chutes for injecting refuse into the boiler	
Traveling grates, 2–One per boiler	Detroit Stoker Co.
Boilers, 2	Babcock & Wilcox Ltd (Can)
Each 106,000 lb/hr at 250 psig (saturated); gas temperature at economizer outlet, 590 F, efficiency, 71%; feedwater temperature, 227 F; excess air leaving the boiler, 37%	
Electrostatic precipitators, 4	Babcock & Wilcox Ltd (Can)
All Lurgi design; two in series per boiler; maximum particulate emissions, 0.08 lb/1000 lb of dry flue gas at 50% excess air.	

Table 13, continued

Thermal Transfer Plant, Nashville, TN	
Owner	Nashville Thermal Transfer Corp.
Consulting engineer	I C Thomasson & Associates Inc.
Commercial operation	April 1974
Incineration capacity	720 tpd
Stoker/grate systems, 2–One per boiler	Detroit Stoker Co.
Boilers, 2	Babcock & Wilcox Co.
Each 135,000 lb/hr at 400 psig/600 F; gas temperature at economizer outlet, 537 F; efficiency, 67.3%; feedwater temperature, 240 F; excess air leaving the boiler, 84%; air-pollution control device, wet scrubber	
Salvage Fuel Boiler Plant, Norfolk, VA	
Owner	U.S. Navy
Consulting engineer	Metcalf & Eddy Construction Engineers
Commercial operation	January 1967
Incineration capacity	360 tpd
Refuse-pit crane–3.5 ton, 2-cu yd capacity	Harnischfeger Corp.
Bulky-refuse shredder	Jeffrey Manufacturing Co.
Motor-driven horizontal hammermill, 30-tph capacity	
Stoker/grate systems, 2–One per boiler	Detroit Stoker Co.
Boilers, 2	Foster-Wheeler Corp.
Each 50,000 lb/hr at 275 psig (saturated). Approximate gas temperatures, F; furnace exit, 1680; boiler outlet, 580. Auxiliary oil burner capacity, 50,000 lb/hr. Sootblowers; one retractable unit between furnace slag screen and boiler bank, two fixed-position units in boiler bank	
Dust collectors, 2	Research-Cottrell Inc.
Each (one per boiler) is involute multiple-cyclone type, limits dust loading to 0.85 lb/1000 lb of dry flue gas adjusted to 50% excess air	
Electrostatic precipitators, 2	Research-Cottrell Inc.
Drag conveyors for bottom ash and flyash	Beaumont Birch Co.
One each	
Refuse Disposal Boiler Plant, Portsmouth, VA	
Owner	U.S. Navy
Consulting engineer	Day & Zimmerman Inc.
Expected startup	Fall 1975
Incineration capacity	160 tpd
Stoker/grate systems, 2–One per boiler	Detroit Stoker Co.
Boilers, 2	E. Keeler Co.
Each 30,000 lb/hr at 115 psig (saturated); auxiliary oil burner mounted in furnace; air-pollution control device, electrostatic precipitator	

Table 13, continued

Thermal Waste Conversion Station, Braintree, MA

Owner	Town of Braintree
Consulting engineer	Camp Dresser & McKee Inc.
Commercial operation	Spring 1971
Incineration capacity	240 tpd
Refuse-pit capacity	1340 cu yd
Refuse-pit crane–5-ton, 3-cu yd capacity	Harnischfeger Corp.
Stoker/grate systems, 2–One per boiler	Riley Stoker Corp.
Boilers, 2	Riley Stoker Corp.
Each 30,000 lb/hr at 250 psig (saturated), auxiliary gas burner	
Electrostatic precipitators, 2	Wheelabrator–Frye Inc.
Each (one per boiler) 93% efficient based on inlet dust loading of 5 lb/1000 lb of dry flue gas and 32,000 cfm (600 F)	

Harrisburg Incinerator, Harrisburg, PA

Owner	City of Harrisburg
Consulting engineer	Gannett, Fleming, Corddry & Carpenter Inc.
Commercial operation	October 1972
Incineration capacity	720 tpd at 5000 Btu/lb
Refuse-pit size	30 ft X 115 ft X 35 ft deep
Refuse-pit cranes, 2–Each 5.5-ton capacity	Dresser Industries Inc.
Crane grapples, 2–Each 3-cu yd capacity	Detroit Stoker Co.
Bulky-refuse shredder	Hammermills Inc.
60-in.-diam. wheel X 80-in. rotor	
Refuse-shredder turbine drive–2000 hp	Elliott Co.
Stoker/grate systems, 2–One per boiler	Josef Martin GmbH (W Ger)
Boilers, 2	Walther GmbH (W Ger)
Each 92,500 lb/hr (continuous) and 120,000 lb/hr (peak) at 250 psig/456 F. Heating surfaces, sq ft: furnace waterwalls, 4236; superheater, 500; evaporator, 15,540; economizer, 6404. Approximate gas temperatures, F: furnace, 1800; furnace exit, 1600; evaporator inlet, 1500; evaporator outlet, 800; economizer outlet, 500. Two auxiliary fuel-oil burners. Forced-draft fan: 62,500 acfm at 25 in. H_2O, 300-hp motor drive. Induced-draft fan: 122,500 acfm at 7 in. H_2O, 200-hp motor drive. Overfire-air fan: 16,800 acfm at 21 in. H_2O, 100-hp motor drive.	
Electrostatic precipitators, 2	Rothemuehle-Walther GmbH (W Ger)
Each (one per boiler) 95% efficient based on an inlet dust loading of 3.5 lb/1000 lb of dry flue gas corrected to 50% excess air	
Water-bath ash dischargers, 2	Josef Martin GmbH (W Ger)
Each (one per boiler) uses 75 gal of water/ton of dry ash	
Belt conveyors for ash, 2–Each 48 in. wide	Jeffrey Manufacturing Co.

Table 13, continued

Bucket elevators for ash, 2–Each 36 in. wide	Jeffrey Manufacturing Co.
Magnetic-metal separators, 2–Each 42 in. wide	Eriez Magnetics
Air-cooled condensers	Marley Co.
Northwest Incinerator, Chicago, IL	
Owner	City of Chicago
Consulting engineer	Metcalf & Eddy Construction Engineers
Commercial operation	Fall 1970
Incineration capacity	1600 tpd at 5000 Btu/lb
Refuse-pit capacity	10,000 cu yd
Refuse-pit cranes, 3–Each 8.5-ton, 5-cu yd capacity	Harnischfeger Corp.
Bulky-refuse shredder	Jeffrey Manufacturing Co.
Horizontal hammermill type, 1200-hp turbine drive	
Stoker/grate systems, 4–One per boiler	Josef Martin GmbH (W Ger)
Boilers, 4	Walther GmbH (W Ger)
Each 110,000 lb/hr (continuous) and 135,000 lb/hr (peak) at 275 psig/414 F; grate combustion efficiency, 95.5%; furnace/boiler efficiency, 66.1%; capacity of two auxiliary gas burners, 110,000 lb/hr	
Electrostatic precipitators, 4	Rothemuehle-Walther GmbH (W Ger)
Each (one per boiler) 96.9% efficient based on an inlet dust loading of 1.6 grains/scf (dry)	
Water-bath ash dischargers, 4	Josef Martin GmbH (W Ger)
Each (one per boiler) produces ash with less than 15% moisture	
Pan conveyors for ash, 2	Jeffrey Manufacturing Co.
Incinerator-3, Montreal, Que., Canada	
Owner	City of Montreal
Commercial operation	Early 1970
Incineration capacity	1200 tpd
Refuse-pit size	180 ft X 30 ft X 60 ft deep
Refuse-pit cranes, 2–Each 7.5-ton, 4-cu yd capacity	Dominion Bridge Co (Can)
Bulky-refuse crusher	Von Roll AG (Switz)
Vibrating feeders, 4	Schenck AG (W Ger)
Each (one per boiler) 11.5 ft X 8.2 ft with 8-deg slope	
Stoker/grate systems, 4–One per boiler	Von Roll AG (Switz)
Boilers, 4	Dominion Bridge Co (Can)
Each 100,000 lb/hr at 225 psig/500 F. Heating surfaces, sq ft: furnace water-walls and evaporator, 15,350; superheater, 1790; air heater, 6080; economizer 2360. Auxiliary oil-burner capacity, 50,000 lb/hr. Forced-draft fan: 55,000 acfm at 22 in. H_2O, 200-hp motor drive. Induced-draft fan: 124,000 acfm (570-F flue gas) at 5 in. H_2O, 200-hp motor and turbine drive on same shaft (only one used)	

Table 13, continued

Electrostatic precipitators, 4	Research-Cottrell Ltd (Can)
Each (one per boiler) 95% efficient; capacity 120,000 acfm at 536 F	
Drag conveyors for ash, 2	Dominion Bridge Co.
Trommel screens, 2–Each 6 ft diam X 10.3 ft long	Beaumont Birch Co.
Air-cooled condensers, 7	Trane Co.
There are four high-pressure condensers–three are rated 31,800 lb/hr each and one, 52,200 lb/hr–and three low-pressure condensers–one rated 15,400 lb/hr; one, 12,800 lb/hr and one, 23,000 lb/hr. Condensate coming from the high-pressure condensers is discharged by steam traps into a flash tank at 15 psig and then transferred to a condensate tank	
North Shore Facility, Saugus, MA	
Owner	Refuse Energy Systems Co.–a joint venture of Wheelabrator-Frye Inc. and M. DeMatteo Construction Co.
Consulting engineer	Rust Engineering Co.
Expected startup	July 1975
Incineration capacity	1200 tpd
Refuse-pit size	39 ft X 200 ft X 85 ft deep
Refuse-pit cranes, 2–Each 13-ton capacity	Harnischfeger Corp.
Bulky-refuse shredder	Hammermills Inc.
Horizontal hammermill type with reversible force-feed system; 25-tph capacity	
Stoker/grate systems, 2–One per boiler	Von Roll AG (Switz)
Boilers, 2	Dominion Bridge Co. (Can)
Each 185,000 lb/hr at 890 psig/875 F based on 750 tpd of refuse with a lower heating value of 4500 Btu/lb; capacity of three auxiliary oil burners, 200,000 lb/hr	
Electrostatic precipitators, 2	Wheelabrator-Frye Inc.
Each (one per boiler) 97.5% efficient; expected particulate emissions, 0.025 grains/scf (0.05 grains/scf required by law)	
Drag conveyors for ash, 2–Each submerged type	Rust Engineering Co.
Trommel screens, 2	Beaumont Birch Co.
Magnetic-metal separators, 2	Eriez Magnetics
Vibrating conveyors for ash, 2	Rexnord, Carrier Div.
Belt conveyors for ferrous scrap, 2	Rust Engineering Co.
Incinerator Plant, Quebec City, Que., Canada	
Owner	Quebec Urban Community
Consulting engineer	Surveyer, Nenniger & Chenevert
Commercial operation	Fall 1974
Incineration capacity	1000 tpd
Stoker/grate systems, 2–One per boiler	Von Roll AG (Switz)
Boilers, 2–Each 81,000 lb/hr at 680 psig/600 F	Dominion Bridge Co. (Can)
Electrostatic precipitators, 2–One per boiler	Research-Cottrell Ltd (Can)

REFERENCES

1. Astrom, L. *et al.* Comparative Study of European and North American Steam Producing Incinerators. *Proceedings, 1974 National Incinerator Conference.* Miami. May 12-15, 1974. American Society of Mechanical Engineers. Pages 225-226.
2. Solid Waste As Fuel For Power Plants. Horner and Shifrin, Incorporated, St. Louis, Missouri. Report EPA-SW-36D-73. U.S. Environmental Protection Agency, Office of Solid Waste Management Programs, NTIS Report PB 220 316. Springfield, Va. 1973. 158 pages.
3. Compiled from RECON SYSTEMS, INC., Princeton, N. J. visits and communiques with operating and planned facilities.
4. Pepperman, C. M. The Harrisburg Incinerator: A Systems Approach. *Proceedings, 1974 National Incinerator Conference.* Miami. May 12-15, 1974. American Society of Mechanical Engineers. Pages 247-254.
5. Bender, R. J. Steam-Generating Incinerators Show Gain. *Power.* McGraw Hill. September 1970. Pages 35-37.
6. Chansky, S. *et al.* Study of Waste Automotive Lubricating Oil as an Auxiliary Fuel to Improve the Municipal Incinerator Combustion Process. Contractor-GCA Corporation (Bedford, Mass.). EPA Contract No. 68-01-0186. Office of Research and Monitoring. U.S. Environmental Protection Agency. U.S Government Printing Office. Washington, D.C. September 1973.
7. Weinstein, N.J. Waste Oil Recycling and Disposal. Contractor-RECON SYSTEMS, INC. (Princeton, N.J.). EPA 670/2-74-052. U.S. Environmental Protection Agency. NTIS Report PB-236 148 Springfield, Va. August 1974. 327 pages.
8. Wilson, M. J. A Chronology of the Nashville, Tennessee Incinerator With Heat Recovery and the Compatible Central Heating and Cooling Facility. *Proceedings, 1974 National Incinerator Conference.* Miami. May 12-15, 1974. American Society of Mechanical Engineers. Pages 213-221.
9. Nashville Turns Solid Waste Into District Steam and Chilled Water. *Power.* December 1974. Pages 18-19.
10. Chapman, R. A. and Wocasek, F. R. CPU-400 Solid-Waste-Fired Gas Turbine Development. *Proceedings, 1974 National Incinerator Conference.* Miami. May 12-15, 1974. American Society of Mechanical Engineers. Pages 347-358.
11. Stabenow, G. Performance of the New Chicago Northwest Incinerator. *Proceedings, 1972 National Incinerator Conference.* New York. June 4-7, 1972. American Society of Mechanical Engineers. Pages 178-194.
12. Mullen, J. F. Steam Generation From Solid Wastes: The Connecticut Rationale Related to the St. Louis Experience. *Proceedings, 1974 National Incinerator Conference.* Miami. May 12-15, 1974. American Society of Mechanical Engineers. Pages 191-202.
13. Regan, J. W. *et al.* Suspension Firing of Solid Waste Fuels. Presented at American Power Conference, Chicago, Illinois, April 22-24, 1969. 7 pages.

14. Klumb, D. L. Solid Waste Prototype For Recovery of Utility Fuel and Other Resources. Technical Paper APCA74-94. Air Pollution Control Association. Pittsburgh, Pa. 1974 Annual Meeting-Denver. 16 pages.
15. Schwieger, R. G. Power From Waste. *Power*. February 1975. Pages S-1 to S-24.
16. Sutin, G. L. The East Hamilton Solid Waste Reduction Unit. *Engineering Digest* **15** (No. 7): 47-51. August 1969.
17. Third Report to Congress. Resource Recovery and Waste Reduction. SW-161. Office of Solid Waste Management Programs. U.S. Environmental Protection Agency. 1975. 96 pages.

CHAPTER 3

PYROLYSIS

Pyrolysis or "destructive distillation" is a process in which organic material is decomposed at elevated temperature in either an oxygen-free or low-oxygen atmosphere. Unlike incineration, which is inherently a highly exothermic combustion reaction with air, pyrolysis requires the application of heat, either indirectly or by partial oxidation or other reactions occurring in the pyrolysis reactor. Again unlike incineration, which produces primarily carbon dioxide and water, the products of pyrolysis are normally a complex mixture of primarily combustible gases, liquids and solid residues. Thus, pyrolysis produces products which are potentially useful as fuels and chemical raw materials.

Several pyrolysis processes have been developed for municipal solid wastes. One full-scale plant is in its early phases of operation in the City of Baltimore, and one is under construction in the County of San Diego. If operations are successful, pyrolysis will be available in the near future as an alternative to incineration and other methods of solid waste disposal.

CHEMISTRY OF PYROLYSIS

The organic portion of municipal solid waste is primarily composed of the elements carbon, hydrogen and oxygen, with minor quantities of nitrogen, sulfur and others. Since the ratios of the major elements approximate those in cellulose, municipal solid waste is sometimes represented chemically as $(C_6H_{10}O_5)_n$, where "n" represents a variable number of the basic chemical units. Indeed, cellulose is a major constituent of solid waste; for example, paper is primarily cellulose, wood contains about 55-60% cellulose, and cotton greater than 90%. For the purposes of this discussion, the chain or polymeric nature of cellulose will be ignored, using the chemical representation $C_6H_{10}O_5$.

Simple Pyrolysis

A simple pyrolysis reaction may be represented by:

$$C_6H_{10}O_5 \rightarrow \text{fuel gas (including some } CO_2 + H_2O\text{) + pyrolytic oil}$$
$$\text{+ other condensibles (oxygenated organics in water)}$$
$$\text{+ carbonaceous solid residue}$$

The relative quantities of gaseous, liquid and solid products and their compositions depend upon the composition of the waste and the conditions of pyrolysis. For example, higher pyrolysis temperature increases gaseous yields. Pyrolysis temperature for processes producing high yields of pyrolytic oil would be about 500°C (932°F), while processes producing primarily gaseous fuels will most likely attain 700-1000°C (1292-1832°F). Solid residues are produced in either case.

Other Pyrolysis Reactions

Product yields can also be shifted by the application of catalysts, high pressure, by the use of oxidizing reactants such as air, oxygen or water, or by the use of reducing reactants such as hydrogen or carbon monoxide. For example, the following types of reactions are possible:

$$C_6H_{10}O_5 + \tfrac{1}{2}O_2 \rightarrow 6CO + 5H_2 \text{ (partial oxidation)}$$
$$C_6H_{10}O_5 + H_2O \xrightarrow{\text{heat}} 6CO + 6H_2 \text{ (reforming)}$$
$$C_6H_{10}O_5 + H_2 \xrightarrow{\text{pressure}} \text{oil} + H_2O \text{ (hydrogenation)}$$
$$C_6H_{10}O_5 + 12H_2 \xrightarrow{\text{pressure}} 6CH_4 + 5H_2O \text{ (hydrogasification)}$$
$$C_6H_{10}O_5 + CO + H_2O \xrightarrow{\text{pressure}} \text{pyrolytic oil (hydro-oxynation)}$$

Systems in which the partial oxidation and reforming reactions predominate would be expected to produce primarily fuel gas, with considerable oxygenated liquids under low-temperature conditions.[1] Hydrogenation reactions may lead to oil (300-350°C, 200-300 atm)[2] or to methane (650°C, 80-200 atm).[3] Hydro-oxynation, at 350°C and high pressure, yields pyrolytic oil.[4] Carbonaceous residue is produced in all cases. Only simple pyrolysis and partial oxidation are practiced in the pyrolysis systems approaching commercialization.

The composition of fuel gases produced also depends upon pyrolysis conditions. Where air is introduced for partial oxidation, the fuel gas is diluted with N_2, limiting its use to industrial equipment especially designed for low volumetric heating value gases. Hydrogenation processes producing CH_4 result in higher heating value gases which may approach natural gas, depending mainly on the presence of unreacted H_2 which has an acceptable but lower volumetric heating value than CH_4. Gases containing CO/H_2 mixtures can be converted to natural gas substitutes.

Pyrolysis Product Compositions

A simple laboratory pyrolysis of dried shredded municipal wastes, with most of the inorganics removed, at about 500°C and atmospheric pressure resulted in the products shown in Table 14.[5]

Table 14. Simple Pyrolysis

Fraction	Char	Pyrolytic Oil
Yield, weight %	20	40
Composition, weight %		
Carbon	48.8	57.5
Hydrogen	3.9	7.6
Nitrogen	1.1	0.9
Sulfur	0.3	0.1
Ash	31.8	0.2
Chlorine	0.2	0.3
Oxygen (by diff.)	13.9	33.4
	100.0	100.0
Heating Value, cal/g (BTU/lb)	5000 (9000)	5830 (10,500)
Fraction	**Gas**	**Water**
Yield, weight %	27	13
	Volume %	**Contains**
	0.1 Water	Acetaldehyde
	42.0 Carbon Monoxide	Acetone
		Formic Acid
	27.0 Carbon Dioxide	Furfural
	10.5 Hydrogen	Methanol
	0.1 Methyl Chloride	Methyl Furfural
	5.9 Methane	Phenol
	4.5 Ethane	Etc.
	8.9 C_3 to C_7 hydrocarbons	
	99.0	
Gross Heating Value, Kcal/NCM (BTU/SCF)	5172 (550)	

The advantage for removal of inorganics from solid waste before pyrolysis can be inferred from the product compositions presented in Table 14. For example, even with prior removal of inorganics, the char produced contained a very high ash content, making this product only marginally useful. If inorganics such as metals and glass are not removed prior to pyrolysis, the higher ash content would most likely relegate the char to the status of a waste product. Therefore, resource recovery prior to pyrolysis is doubly advantageous.

Drying of wastes prior to pyrolysis is also advantageous. Condensation of water formed during pyrolysis produced the water fraction indicated in Table 14. If the waste had not been dried, this fraction would be even larger, diluting the soluble organics produced during pyrolysis and making more difficult their recovery or disposal. Feed moisture also adds to the amount of heat which must be added to the pyrolysis reactor.

As indicated earlier, higher pyrolysis temperature increases the amount of gaseous product. For example, for the combustible portion of a solid waste containing 19.77% free moisture, yields measured in the laboratory are shown in Table 15.[6] Gas composition also varies with pyrolysis

Table 15. The Effect of Temperature on Pyrolysis Yields

Pyrolysis Temperature, °C (°F)	482 (900)	649 (1200)	816 (1500)	927 (1700)
Product Yields, weight %				
Gases	12.33	18.64	23.69	24.36
Volatile Condensibles[a]	43.37	49.20	47.99	46.96
Other Condensibles	17.71	9.98	11.68	11.74
Char	24.71	21.80	17.24	17.67
	98.12	99.62	100.60	100.73

[a]Portion of condensibles which evaporate at 103°C, including water.

temperature, with Table 16 showing hydrogen content increasing as temperature increases.[6] On the other hand, liquid composition does not change drastically with temperature, as shown in Table 17. An additional 33 organic compounds were identified, but all were present in concentrations of less than about 0.3%.

As could be expected, char analyses in Table 18 show decreased volatile matter as pyrolysis temperature increases.[6]

Table 16. The Effect of Pyrolysis Temperature on Gas Composition

Temperature, °C (°F)	482 (900)	649 (1200)	816 (1500)	927 (1700)
Gas Composition, volume %				
Carbon Monoxide	33.50	30.49	34.12	35.25
Carbon Dioxide	44.77	31.78	20.59	18.31
Hydrogen	5.56	16.58	28.55	32.48
Methane	12.43	15.91	13.73	10.45
Ethane	3.03	3.06	0.77	1.07
Ethylene	0.45	2.18	2.24	2.43
	99.74	100.00	100.00	99.99
Heating Value,[a] cal/NCM 2930	2930	3780	3680	3610
(BTU/SCF)	(312)	(403)	(392)	(385)

[a]Gross heating value by calculation.

Table 17. The Effect of Pyrolysis Temperature on Organic Product Composition

Pyrolysis Temperature, °C (°F)	649 (1200)	816 (1500)
Weight % of Condensible Organics		
Acetaldehyde	13.0	10.5
Acetone	18.0	16.5
Methylethylketone	4.3	4.9
Methanol	20.6	23.5
Chloroform	1.0	2.1
Toluene	1.3	3.2
Formic Acid	14.4	11.2
Furfural	7.2	8.0
Acetic Acid	1.3	2.1
Methylfurfural	6.9	6.7
Naphthalene	1.6	1.8
Methylnaphthalene	1.3	1.4
Phenol	6.5	5.6
Cresol	2.6	2.5
	100.0	100.0

Table 18. The Effect of Pyrolysis Temperature on Char Composition

Pyrolysis Temperature, °C	482	649	816	927
(°F)	(900)	(1200)	(1500)	(1700)
Char Composition, weight %				
Volatile Matter	21.81	15.05	8.13	8.30
Fixed Carbon	70.48	70.67	79.05	77.23
Ash	7.71	14.28	12.82	14.47
	100.00	100.00	100.00	100.00
Gross Heating Value, cal/g	6730	6840	6400	6330
(BTU/lb)	(12,120)	(12,280)	(11,540)	(11,400)

PYROLYSIS PROCESSES

Early work in solid waste pyrolysis was naturally analogous to wood and coal pyrolysis or "destructive distillation." These have usually been batch retort or furnace processes with heat applied externally. However, recent developments in solid waste pyrolysis are in the direction of continuous modern engineering technology.

Many pyrolysis process developments have been undertaken in recent years.[7] Those believed to be under active development and to have reached the pilot plant stage on municipal solid waste are summarized in Table 19.

Only the Monsanto, Occidental, Union Carbide, and Carborundum processes are considered sufficiently advanced for further discussion here.

Monsanto Envirochem LANDGARD Process

The Envirochem LANDGARD System encompasses all operations for receiving, handling, shredding and pyrolyzing waste; for quenching and separating the residue; for generating steam from waste heat, and for purifying the off-gases. In the basic pyrolysis process, shredded waste is heated in an oxygen-deficient atmosphere to a temperature high enough to pyrolyze organic matter into gaseous products and a residue consisting mostly of ash, carbon, glass and metal. A flow chart and process description for the LANDGARD plant are shown in Figure 5.

Waste will be received from trucks and transfer trailers at the plant six days per week and metered from two live-bottom hoppers into their respective shredder lines. After shredding, waste is conveyed to a shredded waste storage system, from which it is continuously fed into the kiln.

Table 19. Municipal Solid Waste Pyrolysis Processes

Developer	Products	Pilot Plant Scale	First Commercial Plant
Monsanto Envirochem Systems, Inc., St. Louis, Mo.	Fuel Gas or Steam, Ferrous Metal, Wet Char, Glass Aggregate	35 ton/day	1000 ton/day; still in shakedown as of end of 1975; EPA support[8]
Occidental Research Corp. (formerly Garrett), La Verne, Calif.	Pyrolytic Oil, Char, Glass, Ferrous Metal, Nonferrous Metal, Organics in Condensate	4 ton/day	200 ton/day; startup scheduled for late 1976; EPA support[9]
Union Carbide Corp., New York, N.Y.	Fuel Gas, Slag	200 ton/day	Pilot plant still in operation in late 1975
Carborundum Environmental Systems, Inc., Niagara Falls, N.Y.	Steam (or Fuel Gas), Slag	75 ton/day	200 ton/day commercial plant under construction in Europe (Andco, Inc.)
Battelle Pacific Northwest Laboratories, Richland, Wash.	Steam (or Fuel Gas)	2 ton/day; 150 ton/day demonstration plant under consideration	–
Pyrolytic Systems, Inc., Riverside, Calif.	Fuel Gas or Electric Power	50 ton/day by late 1976	–
DEVCO Management, Inc., New York, N.Y.	Fuel Gas	50 ton/day	–
Pollution Control, Ltd., Copenhagen, Denmark	Fuel Gas	5 ton/day	–
Urban Research & Development Corp., East Granby, Conn.	Slag, Fuel Gas	120 ton/day	–

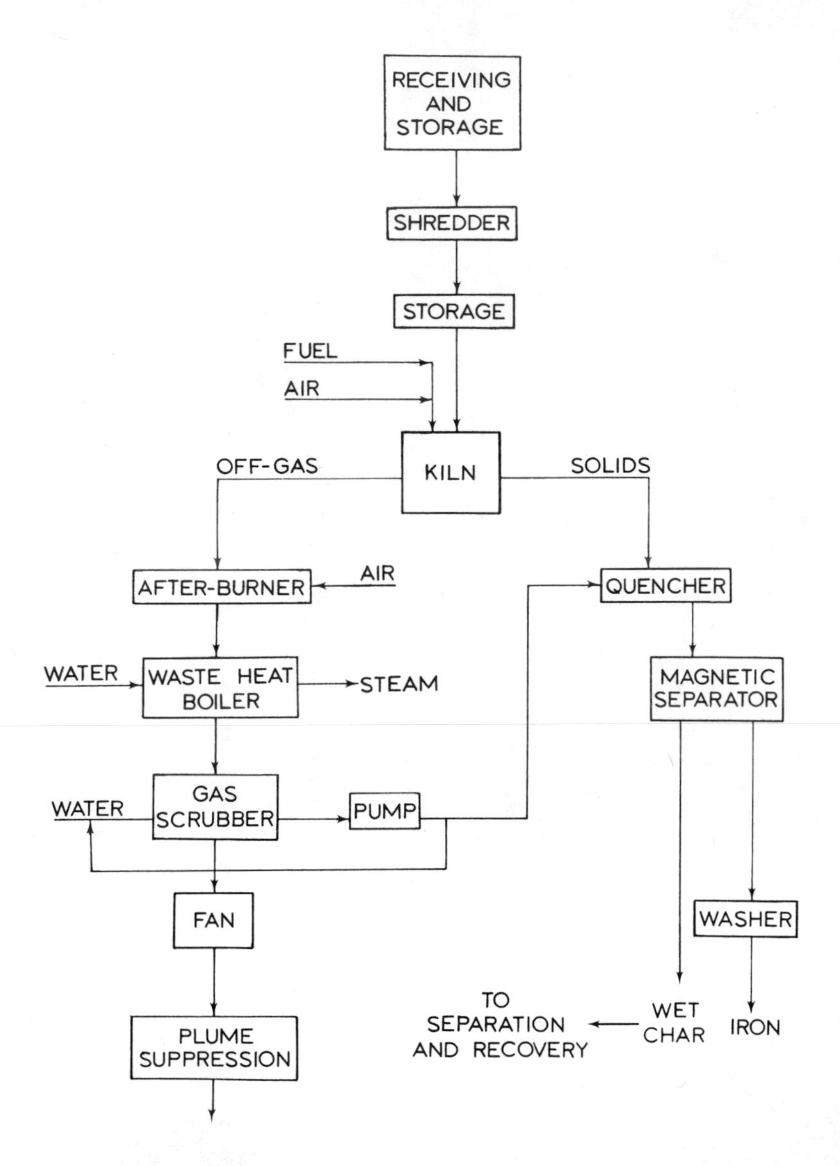

Figure 5. LANDGARD plant flow sheet.

Pyrolysis of shredded waste occurs in a refractory lined horizontal rotary kiln. Shredded waste feed and direct-fire fuel (oil) enter opposite ends of the kiln. Countercurrent flow of gases and solids exposes the feed to progressively higher temperatures as it passes through the kiln so that first drying and then pyrolysis occurs. The finished residue is exposed to the highest temperature, 982°C (1800°F), just before it is discharged from the kiln. The kiln is specially designed (based on successful prototype operation) to uniformly expose solid particles to high temperatures. This maximizes the pyrolysis reaction. The kiln for the Baltimore demonstration plant is designed to handle 38 MT/hr throughput. It is 5.8 m (19 ft) in diameter and 30.5 m (100 ft) long, and rotates at approximately two revolutions per minute.

The hot residue is discharged from the kiln into a water-filled quench tank where a conveyor elevates it into a flotation separator. Light material floats off as a carbon char slurry, is thickened and filtered to remove the water, and conveyed to a storage pile prior to truck transport from the site. Heavy material is conveyed from the bottom of the flotation separator to a magnetic separator for removal of iron. Iron is deposited in a storage area or directly into a railcar or truck. The balance of the heavy material, now called glassy aggregate, passes through screening equipment and then is stored on-site. Plans call for eventual use of the glassy aggregate in "glasphalt" road construction.

Pyrolysis gases are drawn from the kiln into a refractory-lined gas purifier where they are mixed with air and burned. The gas purifier prevents discharge of combustible gases to the atmosphere and subjects the gases to temperatures high enough for destruction of odors.

Hot combustion gases from the gas purifier pass through water tube boilers where heat is exchanged to produce about 2.4 tons of steam per ton of solid waste. Exit gases from the boilers are further cooled and cleaned of particulate matter as they pass through a water spray scrubbing tower.

Scrubbed gases then enter an induced draft fan which provides the motive force for moving the gases through the entire system. Gases exiting the induced draft fan are saturated with water. To suppress formation of a steam plume, the gases are passed through a dehumidifier in which they are cooled (by ambient air) as part of the water is removed and recycled. Cooled process gases are then combined with heated ambient air just prior to discharge from the dehumidifier.

Solids are removed from the scrubber by diverting part of the recirculated water to a thickener. Underflow from the thickener is transferred to the quench tank, while the clarified overflow stream is recycled to the scrubber. Normally all the water leaving this system will be carried out with the residue or evaporated from the scrubber.

Expected stack gas and residue analyses follow in Tables 20 and 21.

Table 20. Stack Gas Analysis

	Volume %
Nitrogen	78.7%
Carbon Dioxide	13.8%
Water Vapor	1.8%
Oxygen	5.7%
Hydrocarbons	<10 ppm
Sulfur Dioxide	<150 ppm
Nitrogen Oxides	<65 ppm
Chlorides	<25 ppm
Particulates	<0.02 grains/SCF dry gas corr. to 12% CO_2

Table 21. Residue Analysis (Wt % Dry Basis)

Proximate Analysis		Ultimate Analysis	
Volatiles	5.5	Metal (Fe)	21.9
Fixed Carbon	12.5	Glass + Ash	60.1
Inerts	82.0		
		Carbon	14.5
		Sulfur	0.1
		Hydrogen	0.5
		Nitrogen	0.2
		Oxygen	2.7

Higher Heating Value = 2500 BTU/lb

pH = 12.0

Water Soluble Solids 2%

Putrescibles < 0.1% (E/C anal. method)

Occidental Process

A simplified flow diagram of the Occidental Research Corp. (formerly Garrett Research & Development Co.) recycling and pyrolysis process is shown in Figure 6. It incorporates the following operations:

1. Primary shredding of a raw refuse to minus two inches.
2. Air classification to remove most of the inorganics, such as glass, metals, dirt and stones, from the organic feed to the pyrolysis reactor.
3. Drying of the air classifier overheads to about 3% moisture.
4. Screening of the dry material to reduce the inorganic content to less than 4% by weight.
5. Recovery of magnetic metals and glass cullett from the classifier underflow.
6. Secondary shredding of the dry organics to about minus 14 mesh.
7. Pyrolysis of the organics.
8. Collection of the pyrolytic products

The first six of these unit operations may be conveniently grouped together as a feed preparation subsection, the primary function of which is to provide a dry, finely divided, and essentially inorganic-free feed to the pyrolysis reactor. An important secondary purpose is to allow the recovery of clean glass and magnetic metals.

The subsystem shown in Figure 7 is designed to recover over 70% of the glass in the refuse. A proprietary froth flotation technique is employed to obtain a sand-sized, mixed color product of better than 99.5% purity. Ferrous metals are recovered magnetically.

Screening the dry, air-classified wastes successively at 0.635 cm (1/4 in.) and 14 mesh can reduce the inorganic content to about 2 wt %. While about 12 to 14% of the organics pass through the screens, these are ultimately returned to the pyrolysis circuit by subsequent glass recovery operations.

The heart of the pyrolysis feed preparation lies in the secondary shredding operation. A finely divided organic feed to the pyrolysis reactor is desirable if high oil yields at atmospheric pressure are to be achieved.

The Occidental flash pyrolysis process involves the rapid heating in a transport reactor of finely shredded organic materials in the absence of air using recycled hot char to supply heat. This technique was developed to maximize liquid fuel yields. Typical yields were shown in Table 14. The gaseous fuel produced and a portion of the char are used on-site for process heat. Some No. 2 fuel oil, used with product oil to quench the process gas stream, is vaporized with uncondensed gas and also burned for process heat.

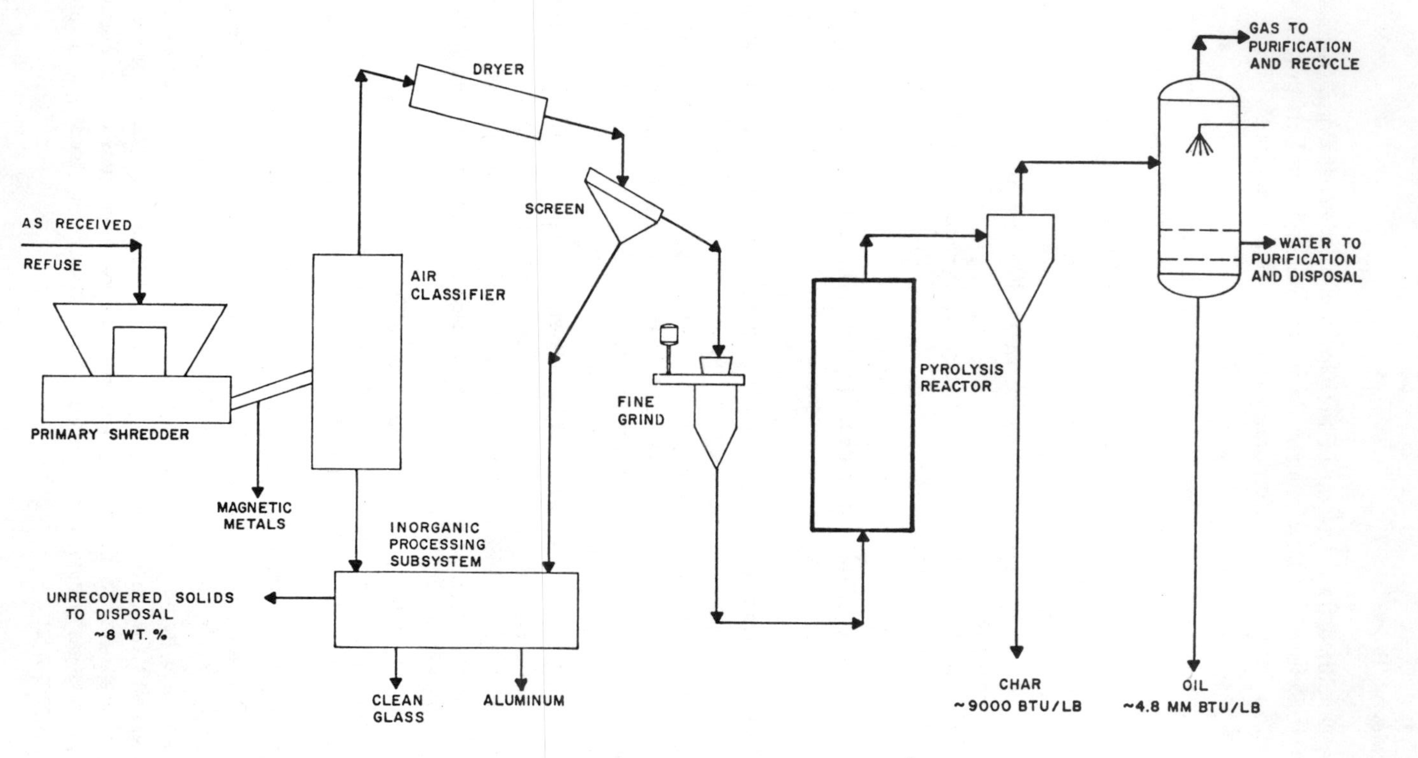

Figure 6. Occidental pyrolysis process.[5]

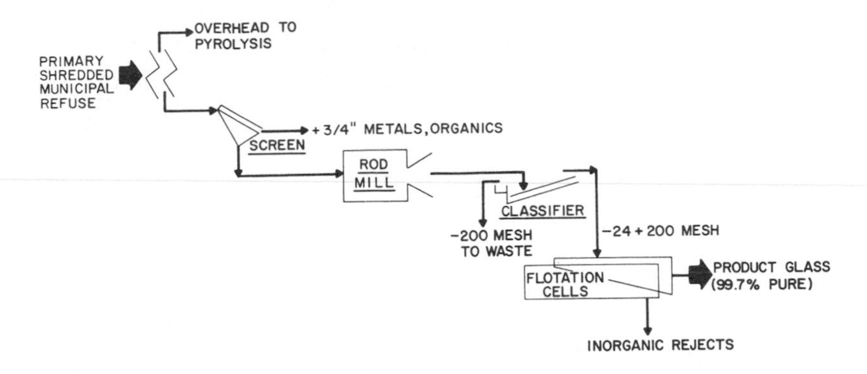

Figure 7. Glass recovery in Occidental pyrolysis process.[5]

Water formed during pyrolysis and condensed from the product gas contains methyl chloride (from polyvinyl chloride pyrolysis) and other organic contaminants such as shown in Table 14. In the San Diego County demonstration project now under construction, these will be oxidized, using fuel gas for heat, in an afterburner (process heater). If markets were available for the energy in the char and fuel gas produced, there would be considerable incentive to develop biological or other treatment or recovery methods to dispose of the contaminated water.

Air streams from the drier, air classifier, and pneumatic transport systems are used as combustion air and passed through the process heater combustion chamber. About 2500 CM/MT (80,000 CF/2000 lb) of hot gas from the process heater (including one-third combustion products) are cooled by preheating various process gas streams, including combustion air for the char heater, and vented through a bag filter for particulate control.

The pyrolytic oil produced is the single most important product. A comparison between typical properties of No. 6 fuel oil and pyrolytic oil is shown in Table 22. Pilot-scale laboratory tests have indicated that the pyrolytic oil can be burned successfully in utility boilers.[9] The San Diego Gas and Electric Company will test and use the pyrolytic oil produced in the San Diego plant at one of its power generating stations. It is expected that at least 0.195 tons of pyrolytic oil will be produced per ton of solid waste (36 gal/ST).[9]

Union Carbide PUROX Process

PUROX is an oxygen-based system to convert municipal refuse into a clean burning fuel gas and a compact, sterile residue. It combines the

Table 22. Typical Properties of No. 6 Fuel Oil and Pyrolytic Oil

	No. 6	Pyrolytic Oil
Carbon, wt %	85.7	57.5
Hydrogen	10.5	7.6
Sulfur	0.7-3.5	0.7-0.3
Chlorine	–	0.3
Ash	0.5	0.2-0.4
Nitrogen	2.0	0.9
Oxygen		33.4
Gross Heat of Combustion, cal/g	10,100	5,800
(BTU/lb)	(18,200)	(10,500)
Specific Gravity	0.98	1.30
Pour Point, °F	65-98	90[a]
Flash Point, °F	150	133[a]
Viscosity SSU @ 190°F	340	1,150[a]
Pumping Temperature, °F	115	160[a]
Atomization Temperature, °F	220	240[a]

[a]Oil containing 14 wt % moisture.

advantages of pyrolysis to produce useful and valuable by-products and high temperatures to melt and fuse the metal and glass. This is made possible by the use of oxygen in the conversion step.

The key element of the system is a vertical shaft furnace (Figure 8). As-received or preprocessed waste is fed into the top of the furnace, and oxygen is injected into the bottom. The oxygen reacts with char formed from the waste. This reaction generates the high temperature in the hearth needed to melt and fuse the metal and glass. This molten mixture drains continuously into a water quench tank where it forms a hard granular material.

The hot gases formed by reaction of the oxygen and char rise up through the descending solid waste and pyrolyze the waste as it cools. In the upper portion of the furnace, the gas is cooled further as it dries the incoming material. This results in the gases exhausting from the furnace at about 93°C (200°F). The exhaust gas contains considerable water vapor, some oil mist, and minor amounts of other undesirable constituents. These components are removed in a gas-cleaning system.

The resultant gas is a clean-burning fuel with about 2821 calories per NCM (300 BTU/SCF) gross heating value (Table 23). It is essentially free of sulfur compounds and nitrogen oxides. It can be effectively used as a supplementary fuel in an existing utility boiler or other fuel consuming operation. The combustion products of this fuel should easily meet air pollution codes.

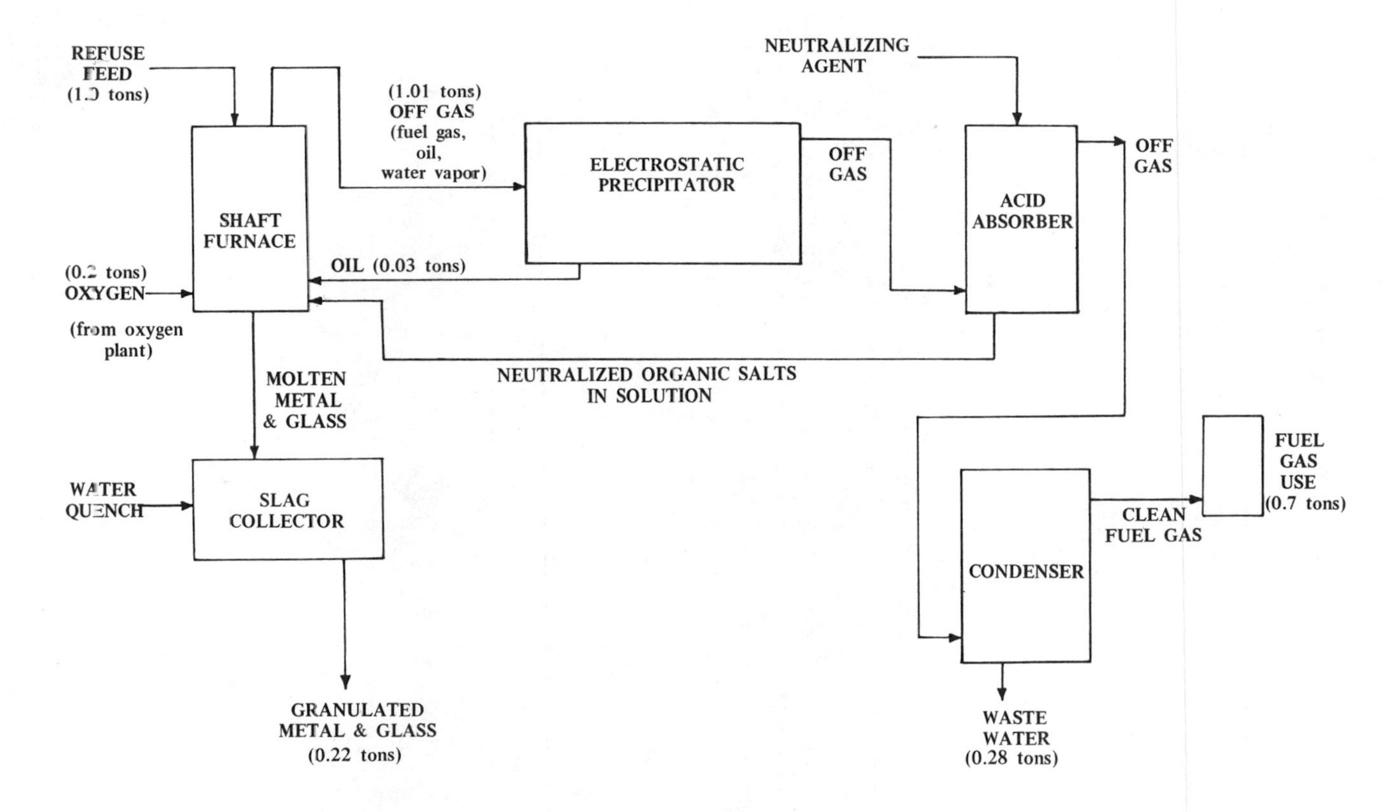

Figure 8. Block diagram–oxygen refuse converter.

Table 23. Fuel Gas Composition

Constituent	Volume %
CO	49
H_2	29
CO_2	15
CH_4	4
C_2H_2 +	1
N_2 + Argon	2
Totals	100

The Union Carbide system is a net producer of energy. The clean-burning fuel gas represents 83% of the fuel value of the original solid waste charged to the conversion system. A minor portion of this fuel gas is used to generate process steam, for building heat, and for the heat energy needed to maintain the auxiliary combustion chamber at operating temperatures. After deducting the aforementioned uses for the fuel gas, approximately 75% of the fuel energy in the municipal solid waste would be available in the remaining fuel gas for other purposes. An energy balance is shown in Table 24.

Table 24. Usage of Available Energy–1000 ST/D Oxygen Refuse Converter Facility

	BTU/hr	Percent
Available Energy in Refuse[a]	416,000,000	100
Energy Losses in Conversion Process[b]	70,000,000	17
Energy Available in Fuel Gas	346,000,000	83
Fuel Gas Uses:		
Process Steam	16,000,000	4
Building Heating	10,000,000	2
Energy to Maintain Auxiliary Combustion Chamber at Operating Temperature	7,000,000	2
Net Energy Available in Fuel Gas	313,000,000	75
Electric Power Generation	30,000 KW[c]	
Electric Power Used in Plant	5,000 KW	
Electric Power Available for Export	25,000 KW	

[a]Based on a refuse heating value of 5000 BTU/lb, this is calculated as (5000)(2000)(1000)/24.

[b]Includes latent heat of moisture in refuse, sensible heat of fuel gas, heat content of molten slag and metal, and heat leak.

[c]Based on combustion of the net fuel gas in a gas utility boiler with an efficiency of 10,433 BTU/KWH (32.7% overall efficiency).

If the fuel gas is used as a supplementary fuel in an existing fossil fired steam boiler, the net energy production is shown in Table 24. For example, the fuel gas from a 1000 ST/day disposal facility could produce 30,000 KW of electric power. Electric power required in the refuse facility is approximately 5000 KW (including power required to separate oxygen from air), resulting in 25,000 KW available for use elsewhere.

The residue produced from the noncombustible portion of the refuse is sterile and compact. Because it has gone through the molten state, it is free of any biologically active material and has been fused to a minimum volume. There is no need to use sanitary landfill techniques for disposal, and it is suitable as a construction fill material. The volume of the residue is 2-3% of the volume of the incoming refuse, depending upon the amount of noncombustible material contained in the refuse. This compares with a residue volume of 5-15% for a conventional incinerator.

An important feature of the PUROX system is that a minimum amount of other materials is introduced and processed with the refuse. This is shown clearly by comparing PUROX with a conventional refractory incinerator. Because a conventional incinerator burns the refuse with an excess amount of air, about seven tons of air are introduced per ton of refuse combusted. This compares with one-fifth of a ton of oxygen introduced per ton of refuse for the PUROX system. This is a 35-fold difference. This difference in input is reflected by a 20-fold difference in volume of gas to be cleaned. This advantage is, of course, offset in part by the cost of separating oxygen from air, or for purchasing oxygen.

Carborundum Torrax System

The Torrax (or Andco-Torrax) System is designed to convert the combustibles in mixed municipal solid waste to a fuel gas by partial oxidation with air, while melting noncombustibles at temperatures up to 1650°C (3000°F). The waste is processed without sorting or pretreatment.

The combustible gas produced is of a relatively low heating value, 1130-1400 Kcal/NCM (120-150 BTU/SCF). The fuel gas can be used in various ways as a source of energy, but, because of its low heating value, it will usually be advantageous to burn the gas and to recover the heat in a waste heat boiler onsite to produce steam for export, or for conversion to electrical power.

The Torrax System consists of the following major subsystems:

1. Gasifier for pyrolysis slagging
2. Secondary combustion chamber to complete oxidation of volatile materials from pyrolysis

3. Regenerative towers for primary combustion air preheating
4. Waste heat boiler to burn fuel gas and convert energy to steam
5. Gas cleaning system

A diagram of the system is provided as Figure 9.

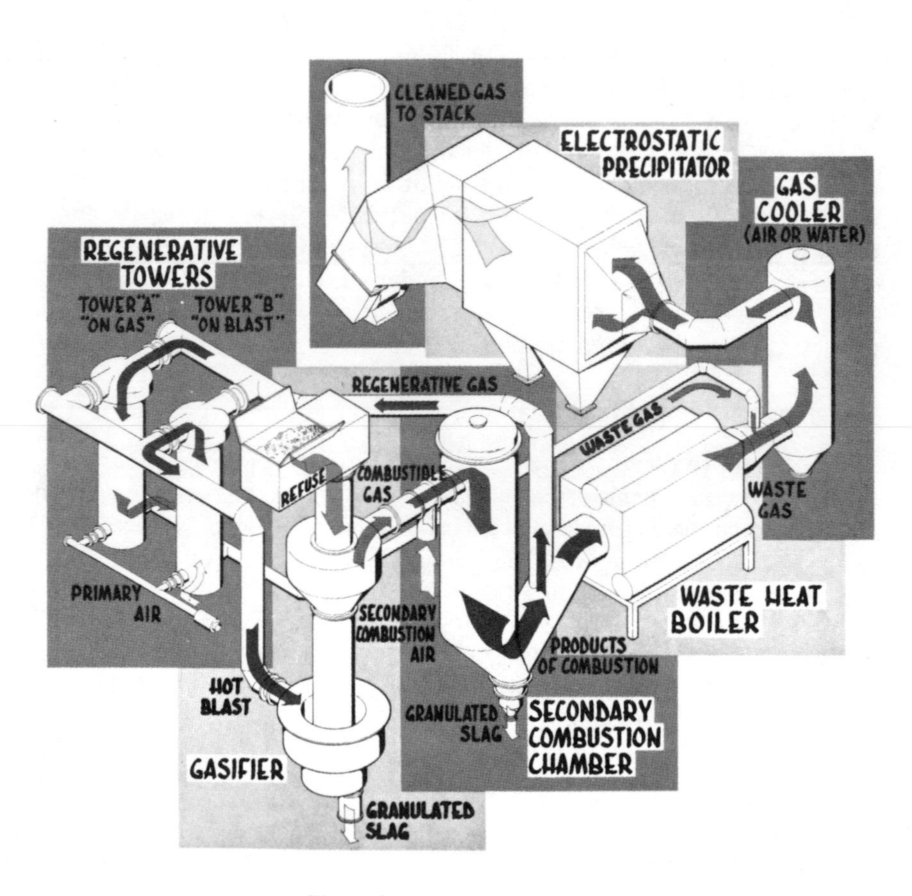

Figure 9. Andco-Torrax System.

Grapple bucket lifts are used to charge waste into the inlet hopper of the gasifier without sorting or pretreatment, except that pieces of refuse larger than one meter in any dimension are sheared before charging. An automatic feeder moves the waste from the hopper into the shaft of the gasifier, as shown in Figure 10. The waste solids then descend by gravity through the drying, pyrolysis, and primary combustion zones of the gasifier. Primary combustion air from the regenerative towers at 1100°C (2000°F) is introduced with some auxiliary fuel at the tuyeres near the base of the gasifier, while the pyrolysis vapors are drawn out of the gasifier in the drying area. The molten slag is drained continuously from

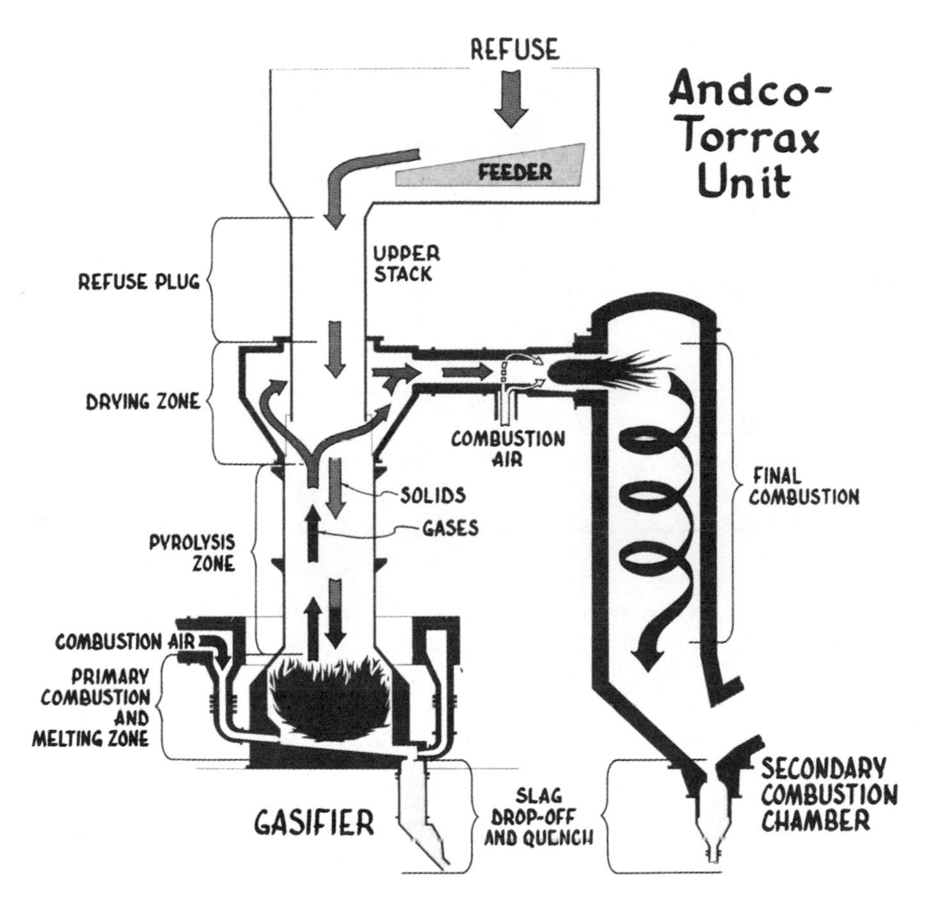

Figure 10. Andco-Torrax unit.

the bottom through a sealed slag tap into a water quench tank to produce a black, glassy aggregate free of carbon or putrescible material. The quantity and composition of the aggregate will vary with the waste.fed, but an example is provided in Table 25. The size of a single gasifier is limited to about 11 MT/hr (300 ST/day).

Table 25. Torrax Aggregate

Constituent	Average % (by weight)	Range %
SiO_2	45	32.0-58.0
Al_2O_3	10	5.5-11.0
TiO_2	0.8	0.48-1.3
Fe_2O_3	10	0.5-22
FeO	15	11.0-21.0
MgO	2	1.8-3.3
CaO	8	4.8-12.1
MnO	0.6	0.2-1
Na_2O	6	4.0-8.6
K_2O	0.7	0.36-1.1
Cr_2O_3	0.5	0.11-1.7
CuO	0.2	0.11-0.28
ZnO	0.1	0.02-0.26
Dry Bulk Density	1.40 g/cc	
True Residue	2.80 g/cc	
Screen Size	4% > 3½ mesh (5.66 mm) 2% < 30 mesh (0.59 mm)	
Quantity, % of municipal solid waste charge	3-5 by volume 15-20 by weight	

The pyrolysis vapors from the gasifier at 450-550°C (800-1000°F) are thoroughly mixed with minimum excess air (10-15%) and burned at 1200-1260°C (2200-2300°F) in the secondary combustion chamber. The high temperature causes flyash and other inert carry-over materials to fuse and be slagged out of the stream. This slag, which is water-quenched, is approximately 10% of the total aggregate produced.

Two refractory-filled steel shells, called regenerative towers, are used to recover heat from about 15% of the hot combustion gas to heat the primary combustion air. These are automatically and alternately controlled to heat the air to 982-1149°C (1800-2100°F) and to cool that portion of the combustion gas used in this subsystem.

The major portion (85%) of the combustion gas from the secondary combustion chamber is cooled in the waste heat boiler to about 260°C (500°F), producing as much as three tons of steam per ton of municipal solid waste.

The waste gases from the regenerative towers and the waste heat boiler are combined, cooled to 300°C (550°F) by water spray or tempering air, and cleaned with a conventional air pollution control system, such as a scrubber or electrostatic precipitator. The normal gaseous emission will contain 81% N_2, 16% CO_2, and 3% O_2 by volume.

Sewage sludge, waste oil, unshredded tires, and polyvinylchloride have been burned with municipal waste in the Torrax pilot unit.

UTILITIES FOR PYROLYSIS PROCESSES

Since little commercial experience is yet available on pyrolysis of municipal solid waste, there is little hard data on utility requirements. The following discussion will serve as an introduction to the expected requirements for the first three processes previously discussed. Insufficient information is available on utility requirements for the Torrax System.

Monsanto Envirochem LANDGARD Process

Power and water requirements for the LANDGARD System, designed to produce steam, may be expected to be similar to those required for incinerators, since complete combustion of offgases with scrubbing and removal of solid residues is practiced. These may approach 70-75 KWH per metric ton of solid waste for power, and about 2-4 tons per ton for water. Fuel requirements are much higher than for incineration, about 0.03 tons of fuel oil per ton (one million BTU per short ton). However, the high cost of utilities in the LANDGARD system is more than offset by the sale of steam, which could also be used internally to decrease electric power consumption.

Occidental Process

The first Occidental plant, designed to produce gaseous and liquid fuels, will not be in operation until late 1976. It is expected to use about 0.0025 tons of fuel oil per ton of solid waste (106,000 BTU/ST), and about 150 KWH/MT of electrical power. Water use will be about 0.35 tons per ton.

Union Carbide PUROX Process

The PUROX Process, which has been operated in a 200 ton per day pilot plant, is designed to produce a fuel gas by partial oxidation with oxygen. The electrical power requirement projected is about 130-140 KWH per metric ton. Most of this power is used for air separation to produce oxygen. The oxygen may be produced on-site, or purchased where available, reducing the actual pyrolysis plant power consumption to a low value. As previously pointed out, the fuel gas produced in the PUROX Process can be used to produce power greatly in excess of that required for oxygen generation and other plant uses.

Fuel and steam requirements in the PUROX Process are equivalent to about 0.02-0.03 tons of fuel oil per ton (0.8 million BTU/ST), much higher than normally required for incineration. Again, this requirement is greatly exceeded by the amount of fuel generated. Water requirement is believed to be similar to that required for incineration.

SUMMARY

Each of the pyrolysis processes discussed will undergo sufficiently large-scale demonstrations, in the period 1976-1977, that the successful processes can be considered as an alternative for municipal solid waste disposal. The obvious advantages of converting solid waste to a valuable energy resource merits a very close examination of this possibility.

REFERENCES

1. Shuster, W. W. Partial Oxidation of Solid Organic Wastes. Final Report. Bureau of Solid Waste Management. SW-7RG. 1970. 111 pages.
2. Groner, R. R., *et al.* The Chemical Transformation of Solid Wastes. AIChE Symposium Series 68(122):28-34. 1972.
3. Feldmann, H. F. Pipeline Gas From Solid Wastes. AIChE Symposium Series 68(122):125-131. 1972.
4. Appell, H. R., *et al.* Hydrogenation of Municipal Solid Waste with Carbon Monoxide and Water. Proceedings of National Industrial Solid Wastes Management Conference. Houston, Texas. March 24-26, 1970. University of Houston and Bureau of Solid Waste Management. Pages 325-379.
5. Mallan, G. M. and Titlow, E. I. Energy and Resource Recovery from Solid Wastes. Occidental Research Corp. (Presented to Washington Academy of Sciences. March 13-14, 1975. College Park, Md.)
6. Pyrolysis of Solid Municipal Wastes. Prepared for National Environmental Research Center, Cincinnati, Ohio. NTIS Report PB 222-015. Springfield, Va. July 1973. 74 pages.

7. Levy, S. J. Pyrolysis of Municipal Solid Waste. Waste Age. October 1974.
8. Sussman, D. A. Baltimore Demonstrates Gas Pyrolysis. First Interim Report. U.S. Environmental Protection Agency. SW-75d.1. Washington, D. C. 1975. 24 pages.
9. Levy, S. J. San Diego County Demonstrates Pyrolysis of Solid Waste. U.S. Environmental Protection Agency. SW-80d.2. Washington, D. C. 1975. 27 pages.

CHAPTER 4

RESOURCE RECOVERY SYSTEMS

Solid waste disposal can be viewed not only as a national problem, but as an opportunity for recovery and utilization of potentially valuable materials. This opportunity can best be exploited by developing economically viable resource recovery projects. Existing and emerging technology, combined with a new conservation awareness by the public, makes imperative the consideration of resource recovery as a part of municipal solid waste management systems.

In its broadest terms, resource recovery from municipal solid waste includes the conversion of solid waste components to energy, fuels and other valuable products, as well as the recovery or salvage of existing materials. Energy recovery via incineration processes and conversion to fuels by pyrolysis have been discussed previously. This chapter will be devoted to recovery of potentially valuable materials as a part of thermal processing systems.

Resource recovery may take place prior to thermal processing, or after the thermal processing step.

RESOURCE RECOVERY PRIOR TO THERMAL PROCESSING

Important materials which, in principle, can be salvaged from raw municipal solid waste include paper products, ferrous metals, nonferrous metals, rags, glass, rubber and plastics. The example in Table 26 for a mixed refuse shows that about 29 wt % could be available for salvage, grossing \$2.27-\$16.10 per metric ton of solid waste. Wide fluctuations in both salvage yield and prices necessitate detailed purity, yield, market and cost studies for each project.

Historically, segregation at the source (*e.g.*, curbside) or handsorting have been the principal approaches to salvage. These are still practiced

Table 26. Potentially Salvageable Materials in a Mixed Municipal Refuse[a]

	Percent by Weight			
	Original Refuse	Potentially Salvageable	Commodity Value, $/MT Material	Potential Revenue $/MT Refuse
Paper and Cardboard	33.0	15.0	5-50	0.75-7.50
Glass	8.0	5.6	5-50	0.28-2.80
Ferrous Metals	7.6	6.8	5-50	0.34-3.40
Nonferrous Metals	0.6	0.6	150-400	0.90-2.40
Plastics, Leather, Rubber, Textiles, Wood	6.4	1.0[b]	–	–
Garbage and Yard Wastes	15.6	–	–	–
Miscellaneous (Ash, Dirt, Etc.)	1.8	–	–	–
	73.0	29.0		$2.27-16.10/MT[c]
Moisture	27.0			($2.06-14.60/ST)
	100.0			

[a]Based in part on data in reference 1.

[b]Potentially recoverable rags and plastics.

[c]Excludes possible credits for energy recovery from unsalvaged refuse.

to a minor extent. The volunteer recycling center is the most recent approach to salvage.

Sorting may be practiced for salvage purposes or to segregate bulky materials, which are then shredded and mixed with the normal municipal solid waste for incineration or disposed of in a landfill. Handsorting is unattractive because of the high cost of labor, and because of the modest degree of separation which is practical, usually limited to bulky items. Separate collection of salvageable materials also tends to be costly, but this approach could become more practical in the future by the development of automated collection and by increasing raw material prices.

Resource recovery prior to thermal processing can be justified not only by the value of the salvaged materials, but also by beneficial effects on the process. For example, removal of noncombustibles such as glass and metals will reduce the ash component of solid and liquid fuels produced during pyrolysis, or improve the incineration process by minimizing furnace damage and decreasing the quantity of solid residue. Removal of noncombustibles is a vital step in the preparation of combustible refuse for use as a fuel component in combination fossil fuel/prepared refuse steam boilers.

Two primary steps are required to practice the type of resource recovery technology now becoming available. The first is size reduction to allow physically freeing from each other the various types of materials present. Thus, size reduction usually precedes the second step, which is physical separation based on utilizing property differences of the materials present. Each of these steps will be discussed in more detail, followed by a discussion of commercially available resource recovery systems.

Size Reduction

The size reduction of municipal solid waste has been called shredding, milling, pulverizing, grinding and comminution, even though only the last term can be considered truly generic. The other terms are more or less related to the type of equipment used. In this publication, the term shredder will be used as a general term for equipment designed to reduce the size of municipal solid waste, except where wet pulpers are used.

There may be as many as 70 suppliers of equipment with the potential for use in municipal waste size reduction.[2,3,6] The many kinds of equipment available are summarized in Table 27. The most common types now in use are the hammermills and rotary ring grinders, but such classifications encompass many possible variations. For example, hammermills may be vertical or horizontal; they may use swing hammers, rigid hammers, and even shredding members; or they may differ in the design of reject systems.

The first stage of size reduction is usually designed to produce material nominally in the 5-25 cm (2-10 in.) range. Ballistic separation of metals may be an integral part of the first stage. A second stage, where required, reduces the size to that required for the process used, or for separation. Classification may be practiced between size reduction stages. A 50 ton per hour shredder will usually require a motor in the 500-1000 horsepower range.

Well designed primary size reduction equipment should be able to handle most objects found in unsorted municipal waste including, for example, home appliances, storage drums, solid wood, and even tires on wheels; but hardened steel objects and flammable and explosive materials can cause severe wear or damage. Other materials, such as rugs, mattresses, wire, or plastic sheets and milk bottles, sometimes cause operability problems. Collection restrictions, presorting, and prescreening can be helpful in a size reduction system, but troublesome materials are difficult to eliminate entirely. Actual tests should be carried out wherever possible before final equipment selection.

Table 27. Current Size-Reduction Equipment and Potential Applications to Municipal Solid Waste[2]

Basic Types	Variations	Potential Application to Municipal Solid Waste
Crushers	Impact	Direct application as a form of hammermill.
	Jaw, roll, and gyrating	As a primary or parallel operation on brittle or friable material.
Cage Disintegrators	Multi-cage or single-cage	As a parallel operation on brittle or friable material.
Shears	Multi-blade or single-blade	As a primary operation on wood or ductile materials.
Shredders, Cutters and Chippers	Pierce-and-tear type	Direct as hammermill with meshing shredding members, or parallel operation on paper and boxboard.
	Cutting type	Parallel on yard waste, paper, boxboard, wood, or plastics.
Rasp Mills and Drum Pulverizers		Direct on moistened municipal solid waste; also as bulky item sorter for parallel line operations.
Disk Mills	Single or multiple disk	Parallel operation on certain municipal solid waste fractions for special recovery treatment.
Wet Pulpers	Single or multiple disk	Second operation on pulpable material.
Hammermills		Direct application or in tandem with other types.

In addition to capital and operating costs, major considerations in the choice of size reduction equipment are durability, reliability, composition of feed, and suitability for the particular separation process contemplated. For example, a wet size reduction system would not be used with a dry separator, nor would a disintegrator producing primarily very fine particles be used with an air classifier. Wear, maintenance, and power input are other major considerations in the choice of size reduction equipment.

Size reduction can be practiced either on the entire waste stream prior to resource recovery, or simply as a method for reducing the size of bulky waste to allow handling in the thermal processing system.

Physical Separation

A thorough investigation into unit processes available for solid waste separation is provided in a 1971 U.S. Environmental Protection Agency report.[2] This report and more recent literature[4,5] discuss the techniques shown in Table 28. However, since solid waste separation is a rapidly evolving technology, the state-of-the-art must be carefully determined at the time of process selection. A few of the more advanced techniques will be described here.

Table 28. Unit Processes for Solid Waste Separation[2,4,5]

	Potential Municipal Solid Waste Application
Magnetic Separation	Magnetic materials (iron)
Inertial Separation Ballistic Secator Inclined Conveyor	Differences in size, density, elastic properties (depending on type)
Eddy-Current Separation	Conductive nonmagnetic materials (copper, aluminum, zinc)
Electrostatic Separation	Aluminum from glass; plastics, paper[19]
Size Classification Vibrating Screens Spiral Classifiers	Preparation for further processing or rough cut separations
Air Classification Vertical Chute Zig-Zag Flow Classifier Horizontal Chute Vibrating Elutriator	Light material, such as paper, from heavier materials
Gravity Separation Dense Media Stoners Tabling Zigging Osborne Dry Separator Fluidized Bed Separator Rising Current Separator	Glass from metals, paper from other materials, and other separations based on density difference
Optical Sorting	Dirt from glass, separation of colored glass
Sweating	Melting to separate metals (*e.g.*, lead and zinc from aluminum)
Flotation	Air bubbles in liquid used to separate materials with differing affinities for air and fluids used
Cryogenic Separation	Difference between materials in tendency to become brittle at low temperature (*e.g.*, liquid nitrogen)

Air Classification

Development studies in recent years have shown the feasibility of air classifying shredded municipal refuse to remove metal, glass, rocks, rubber and wood.[6] As a result, air classification has been incorporated as a primary separation step into several municipal solid waste thermal processing systems, including the preparation of refuse for combined prepared refuse/fossil fuel combustion in steam boilers, for pyrolysis, and for fluidized bed combustion.

In a vertical air classifier, air is drawn upward through a vertical column at a predetermined velocity, while the shredded solids are fed to the top or to an intermediate point. The solid particles are fractionated according to density, size and shape. The particles whose properties are such that they cannot be transported by the airstream move countercurrently to the stream and are discharged at the bottom of the column. The transported particles move with the airstream through a blower and cyclone separator for recovery. In some cases, multiple classifiers in series can be used to separate several different products. Figure 11 is a schematic flow diagram showing one possible arrangement for such a system, while Figure 12 shows a cross section of an air classifier. A supplier's specification for a 45.4 metric ton per hour classifier is provided in Table 29, and an analysis of an air classified light fraction is provided in Table 30.

Ferromagnetic Separation

Magnetic removal of ferrous metals can be practiced prior to air classification, after air classification, or both. As will be discussed later, when resource recovery is not practiced prior to incineration, ferrous metals are sometimes recovered from incinerator residue. Removing ferrous metals from municipal solid wastes is basically rather simple, but problems arise from contamination of the recovered metal with refuse entrapped by the metal as it is attracted to the magnetic separator.

The magnets used can be of a permanent type, or electromagnets. Direct current for the electromagnets normally requires purchase of a rectifier. Alternating current is required for belt drives. When electromagnets are used, designs must allow for lower magnetic strength at operating temperatures as compared to cold start-up temperature.

Rotating drum and suspended magnets have been used for primary separation. These and pulley type magnetic separators can be used after classification.[2,7] Although magnetic separators have been widely used in industry, design changes have been necessary to adapt these to processing of municipal solid waste. For example, multiple magnets in a suspended separator with a moving belt and modified drums have been

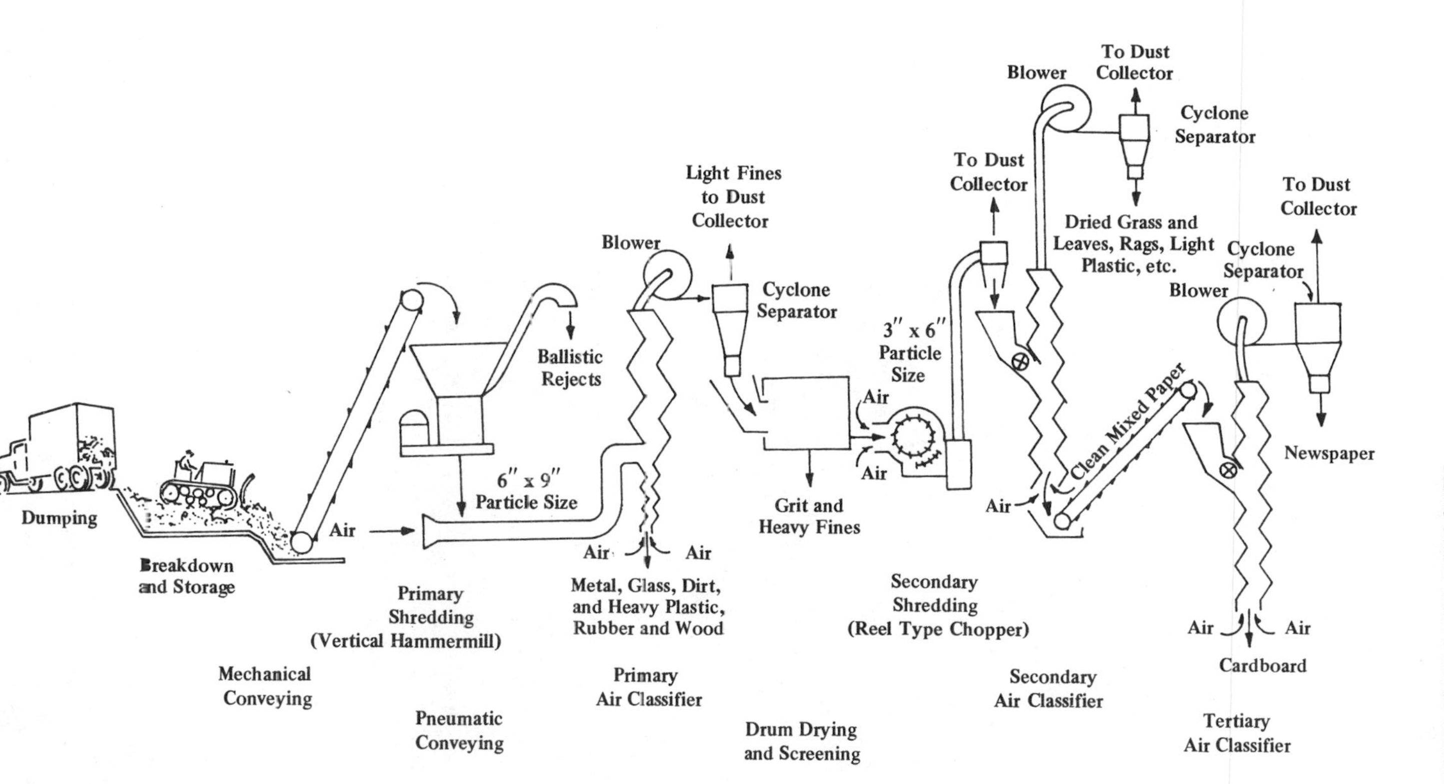

Figure 11. Air classification of solid wastes.[6] Schematic flow diagram of air classification process for wastepaper recovery from municipal refuse. (Secondary and tertiary air classification columns are sufficiently similar that a single column might be used in a time-phase operation by storing the clean mixed paper product and running in the secondary column when no refuse is being received.)

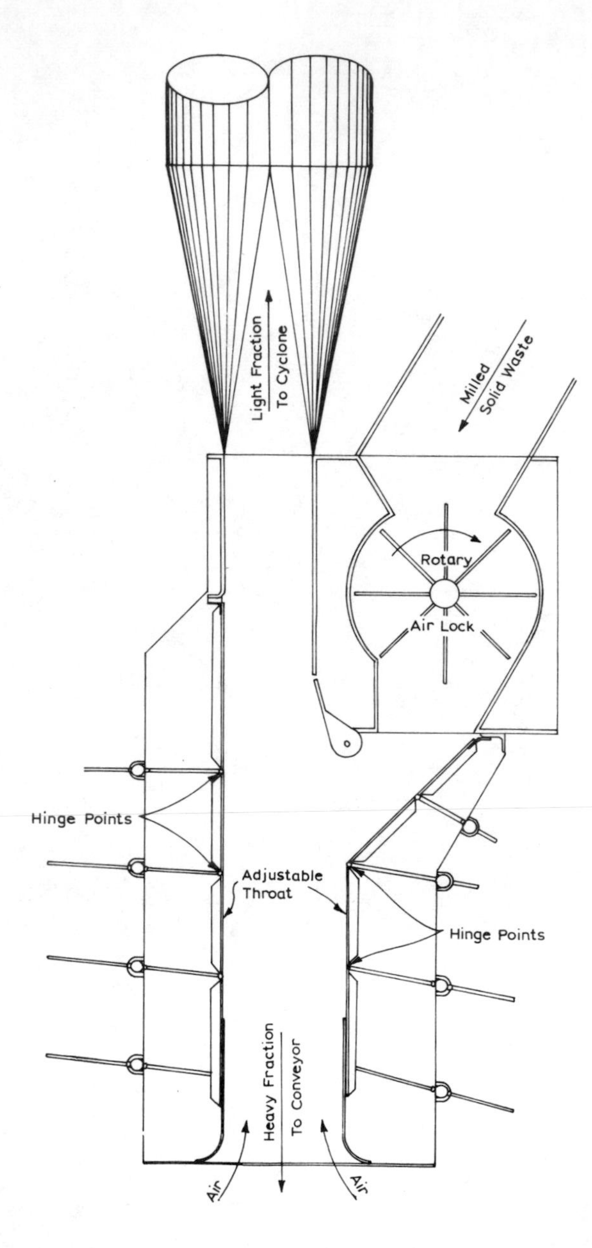

Figure 12. Cross section of an air classifier.[30]

Table 29. Specification for an Air Classifier[9]

Feed: Milled residential, industrial, commercial solid waste
    Particle size - 95% less than 20.23 cm (8 in.)
                    99% less than 30.48 cm (12 in.)
    Moisture - 0-40% by weight
    Density - 0.064 to 0.321 g/cc (4-20 lb/ft^3) loose
Envelope Dimensions (includes conveyors, blower, stands, etc.):
    15.24 meters long by 15.24 meters wide x 12.19 meters high (50 x 50 x 40 ft)
Weight Installed: 54.4 metric tons (60 short tons)
Electrical Requirements: 440 Volts - 3 Phase - 60 Hz
Total Installed Horsepower: 100-200
Nominal Performance:
    Capacity - 41 to 54 metric tons per hr (45-60 short tons/hr)
    Light Product (80-90 weight %) - Less than 10 weight % inerts (fine glass, sand)
    Heavy Product (10-20 weight %) - Less than 5 weight % fibrous material
                                    - 98% of the ferrous metals
                                    - 80% of the aluminum
                                    - 98% of the other metals
                                    - 80% of the glass

developed to separate nonmagnetic entrapped materials by successive attraction and release of the ferrous metals.

Nonferrous Metal Separation

A promising advance in this field is the development of eddy-current techniques for the separation of conducting nonmagnetic materials from municipal solid waste which has been pre-processed by shredding, classifying, and ferrous metal removal.[7] The trash-metal mix from the pre-processing steps is conveyed into a polyphase alternating current electromagnetic field which induces electrical currents in conducting materials, such as aluminum, generating in turn magnetic flux opposite in direction to the initially imposed flux. The resulting repulsive force sweeps aluminum can stock laterally off the belt for collection.

A single aluminum separator module is designed to handle approximately 1.3-2.3 metric tons per hour of trash-metal mix recovered from about 9 metric tons per hour (10 short tons/hr) of shredded waste. Therefore, a 36-metric ton per hour (960 short tons/day) thermal processing plant would require a splitter and at least four modules. Other limitations of this approach include the necessity for a relatively high speed belt to allow spreading out the feed, reducing the probability of

Table 30. Air Classified Refuse Analyses Light Fraction (194 samples taken November 9, 1973 through March 28, 1974[25])

	As Received Basis, Wt %						
				Chlorides			
	Moisture	Ash	Sulfur	Total	NaCl	Cal/g	(BTU/lb)[a]
Average	30.3	16.8	0.10	0.41	0.33	2768	(4983)
Maximum	66.3	31.3	0.28	0.94	0.59	4218	(7593)
Minimum	11.1	7.6	0.04	0.14	0.11	1274	(2293)

	Ash Analysis, Wt %		
	Average	Maximum	Minimum
P_2O_5	1.43	2.04	0.99
SiO_2	49.90	58.10	39.90
Al_2O_3	11.38	26.90	6.10
TiO_2	0.87	1.52	0.07
Fe_2O_3	7.89	22.19	3.03
CaO	12.21	15.80	8.51
MgO	1.29	2.32	00.22
SO_3	1.48	3.75	0.54
K_2O	1.57	2.91	0.92
Na_2O	8.87	19.20	3.11
SnO_2	0.05	0.10	0.02
CuO	0.32	1.74	0.08
ZnO	0.41	2.25	0.09
PbO	0.19	0.73	0.04

[a]Higher heating value

extraneous material being swept off the belt with the aluminum stock; the necessity for careful shredding both to avoid too fine shredding which can cause aluminum flaking and loss in pre-processing steps, and too coarse shredding which can result in poor air classification increasing the contamination and quantity of the trash-metal fraction; and the potentially increased contamination level of recovered aluminum when the fraction of aluminum in the shredded waste is low.

System specifications for a four-module aluminum separator are shown in Table 31. A second-stage separator is under development to recover a mixed product containing the remaining aluminum and other metallics such as copper, zinc, stainless steel and brass.

The heavy fraction from air classification can also be processed with a series of screens and sink-float (dense media) devices to separate individual metals. This approach is under development.

Table 31. Specifications for a Dry Aluminum Separation Technique[7,8]

Modules: Four (for a 36 MT/hr thermal processing plant)

Feed: Dense fraction of milled classified residential, commercial solid waste, less ferrous metals

- Particle size - 95% less than 15.24 cm (6 in.)
- Aluminum size - 90% greater than 2.54 cm (1 in.)
- Moisture - No limitations
- Density - Greater than 0.64 g/cc (40 lb/ft^3)
- Shape - No limitations

Envelope Dimensions (includes conveyors, screen, separators):

12.2 m long by 6.1 m wide by 3.05 m high (40 ft x 20 ft x 10 ft)

Electrical Requirements: 440 Volts - 3 Phase - 60 Hz

Total Installed Horsepower: 75 (equivalent)

Efficiency: 60-80% recovery of can stock material (based on aluminum in dense fraction); about 50% of total aluminum in solid waste

Aluminum Analyses (based on pilot plant samples):

Chemical Analysis, wt %

Element	Sample I	Sample II	Alcoa Grade I
Si	0.28%	0.28%	0.3%
Fe	0.43	0.41	0.5
Cu	0.14	0.16	0.25
Mn	0.84	0.83	1.25
Mg	0.96	0.99	2.0
Cr	0.02	0.02	0.2
Ni	0.00	0.00	0.2
Zn	0.05	0.45	0.5
Ti	0.02	0.02	–
V	0.01	0.01	–
Pb	0.00	0.00	0.1
Sn	0.00	0.00	0.1
Bi	0.00	0.00	0.1

Hand Picked Analysis, wt %

Pieces of Cans or Containers	91.7%
Heavy Material	7.5
Foil	0.1
Dirt	0.7
	100.0

Wet Pulping

The organic and friable fractions of solid wastes can be converted into a water slurry in equipment known as a Hydrapulper.[10] In this pulping process, after separation of nonsuitable feed materials such as tires, large appliances, and building demolition wastes, water is added to the solid waste in a large mixing vessel containing a high-speed cutting rotor, similar to a Waring blender. Nonpulpable material, such as metal cans and stones, are ejected through an opening in the side of the mixing vessel. The slurry, containing about 3-4 wt % solids, is removed through a perforated plate at the bottom to a liquid cyclone and other equipment for subsequent recovery and separation of metals, glass and organics. Wet pulping, in effect, serves both as a size reduction method and as a first step in a series of physical separations.

Usable long paper fibers can be recovered from the slurry for sale as a low-grade paper fiber, leaving an organic residue suitable for thermal processing, or the entire organic fraction can be recovered from thermal processing without separating the paper fibers. A 5-7 metric ton per hour (150 short tons/day) demonstration has been conducted.[10] The Hydrapulper in this operation was 3.66 m (12 ft) in diameter and equipped with a 300-horsepower motor.

Glass Separation

The recovery of mixed glass in a resource recovery operation will normally be from a secondary or tertiary separation step, after separation of light materials, such as paper and metals. The methods used will vary with the overall scheme, involving processes such as air classification, dense media separation, froth flotation, and water elutriation. However, the mixed glass product has limited value unless it is color-sorted and free of contaminants.[11] Since a market does exist for pure color-sorted glass cullet, there is considerable incentive for automatic color sorting to separate flint (clear), amber and green fractions. This technology is under development.[12,13]

In the Sortex optical separator, a continuous stream of individual particles are dropped through an optical box, containing three photocell assemblies set at 120° intervals and suitable illumination sources. Opposite each photocell head is a background with variable shades of color. Each particle passes through the viewing area and, if there is a change in its reflectivity with respect to the background standard, either lighter or darker as desired, a blast of compressed air is triggered to deflect the off-color particle from the main stream. Two optical separators in series will first separate flint (clear) from colored glass, and then separate the colored glass into amber and green.[14]

Conveying Systems

Conveyor performance plays a major role in determining the reliability of resource recovery systems. Various conveyors used are called infeed conveyors, for feeding shredders; transfer conveyors, for transferring shredded material from the shredder discharge to a discharge conveyor; discharge conveyors, for discharging to a storage area or the next processing step; and other conveyors used for magnetic separation, changes in direction, etc.

Belt conveyors are easily maintained and are usually the least expensive of the conveyor types available, but they are subject to failure by impact of sharp objects fed onto the belt from trucks, cranes or other feeders. Their minimum speed of about 24 meters per minute (80 ft/min) for good tracking is generally too fast for feeding a shredder. On the other hand, speeds greater than 24 meters per minute may cause paper and other light materials to float. In addition, angles greater than 20° without cleating are not recommended. Therefore, the best uses for belt conveyors in resource recovery systems are for conveying heavy materials at high speeds over level areas, and for use with magnetic separators where ordinary steel conveyors interfere with magnetic operation.

Apron conveyors are preferred for infeeding and transfer conveyors, even though capital cost, horsepower and maintenance are greater than for other types of conveyors.[15] For infeeding, these conveyors are typically 1.2 to 1.8 meters (48-72 in.) wide and travel 1.5-7.6 meters per minute (5-25 ft/min) with a variable speed drive. They can be inclined to approximately 35° with 10-cm (4-in.) flights welded to the pans. Compression feeders and leveling conveyors are often used with apron conveyors for infeeding to shredders. Special attention must be paid to the details of conveyor construction and to insure proper operation, rugged construction, and ease of maintenance.[15] A typical shredder installation with conveyors is shown in Figure 13.

Integrated Resource Recovery Systems

Although important advances have been made in recent years, no complete resource recovery system can be categorized as fully developed for widespread application. A summary of available systems which can provide feed for thermal processing is provided in Table 32. Flow diagrams for some of these systems are provided as Figures 14-20.

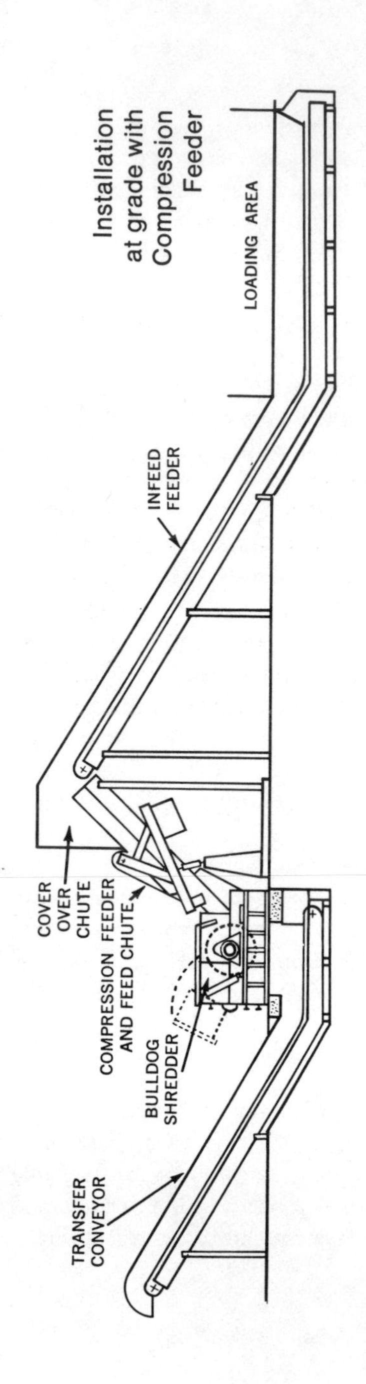

Figure 13. Typical shredding plant.

Table 32. Status of Available Integrated Resource Recovery Systems

Name/Developers/Sponsors	Primary Products	Secondary Products	Development Status[b]
Horner and Shifrin; Union Electric Co.; City of St. Louis, Mo.[a]	Fuel, ferrous metal	Power from fuel in power plant	37 MT/hr (650 ST/D over 2 shifts) evaluation underway[12]
CPU-400, Combustion Power Co.[a]	Fuel, ferrous metal, aluminum, nonferrous metal mix, dirty glass, trash	Power from fuel burned in high-pressure fluidized bed and flue gas discharged to turbines	3 MT/hr (80 ST/D) pilot plant evaluation-resource recovery promising/energy recovery problems to be solved[12]
Hydrasposal, Black Clawson Co.; Franklin, Ohio; Glass Container Manufacturers Institute[a]	Fuel, paper fiber, ferrous metal, color-sorted glass, aluminum	Credit for sewage sludge disposed of in combustor	5.7 MT/hr (150 ST/D) evaluation underway[12]
Bureau of Mines[a]	Fuel, ferrous metal, aluminum nonferrous metal mix, sorted glass	–	4.5 MT/hr (5 ST/hr) pilot plant[13]
Occidental Research Corp.; County of San Diego; San Diego Gas & Electric Co.[a]	Pyrolysis feed, ferrous metal, mixed glass, residue	Pyrolytic oil, high ash char	Pilot plant operations–7.6 MT/hr (200 ST/D) plant under construction, to be completed in late 1976
LANDGARD, Monsanto Enviro-Chem Systems, Inc.; City of Baltimore; State of Maryland[a]	Pyrolysis feed, ferrous metal	Steam, glassy aggregate, char residue	37.8 MT/hr (1000 ST/D) plant in shakedown in late 1975
Hercules, Inc.; State of Delaware[a]	Fuel, ferrous metal, heavy residue (from municipal refuse)	Power from fuel in power plant; pyrolysis of industrial waste with municipal residue to recover fuels, nonferrous metal, glass	Plant to handle 18.9 MT/hr (500 ST/D) municipal refuse plus industrial waste and sewage sludge being designed[12]

Table 32, continued

Eco-Fuel, Combustion Equipment Assoc.	Fuel, glass, metals	–	45 MT/hr (1200 ST/D) capacity plant in East Bridgewater, Mass.–performance unknown[22]
Gibbs, Hill, Durham & Richardson Inc.; City of Ames, Iowa	Fuel, ferrous metal, aluminum, nonferrous metal mix, dirty glass, residue	Power from fuel (in municipally owned power plant)	7.9 MT/hr (210 ST/D) of segregated waste[12,20]; shakedown in late 1975
National Center for Resource Recovery, Inc.; City of New Orleans; Waste Management, Inc.	Fuel, ferrous metal, aluminum, nonferrous metals, sorted glass	Fuel (light combustibles) may be landfilled initially	24.6 MT/hr (650 ST/D) scheduled for startup in summer 1976
State of Connecticut; Occidental Research Corporation; Combustion Equipment Associates; American Metals Climax; SCA Services, Inc.	Fuel, ferrous metal, aluminum, glass, residue	Power from fuels in power plant	Design of systems for Bridgeport, and New Britain/Hartford to total 136 MT/hr (3600 ST/D) scheduled for operation by mid-1976[17]
Sira International Corp.	Fuel, ferrous metal, residue	–	4.5 MT/hr demonstration plant in Los Gatos, Calif. to be expanded to 18 MT/hr, performance unknown; fuel pelletized[23]
Americology, American Can Co.	Fuel, ferrous metal, aluminum, glass, paper	Power from fuel in power plant	45.4 MT/hr (1200 ST/D) plant being designed for Milwaukee, Wisc.[21]
Eastman Kodak	Fuel, metal	Steam from incineration of fuel	11.3 MT/hr (300 ST/D) plant operating in Rochester, N.Y.

[a]Indicates federal support.[12]

[b]MT = metric tons (2205 lb), ST = short tons, D = day.

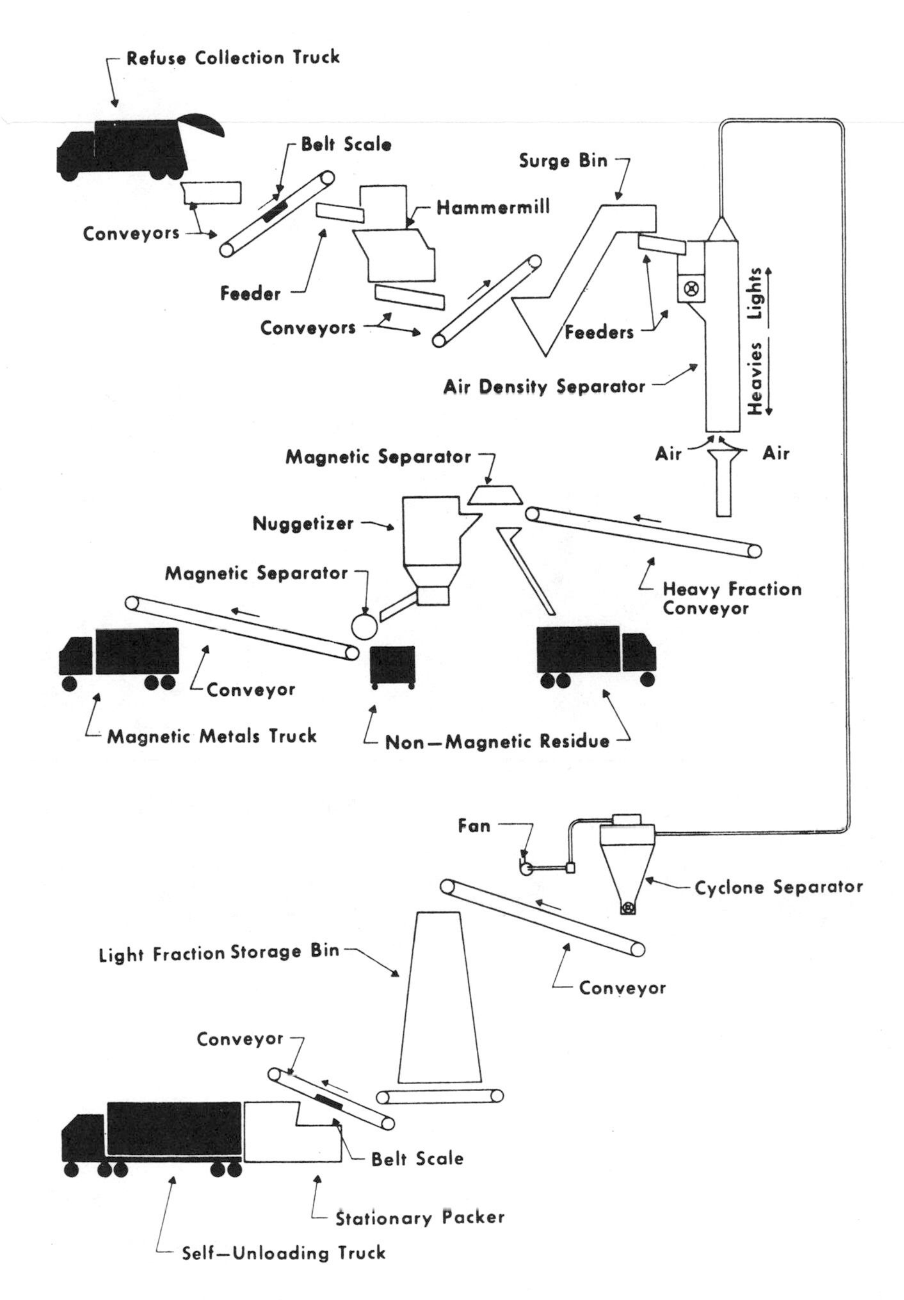

Figure 14. City of St. Louis solid waste processing facilities.

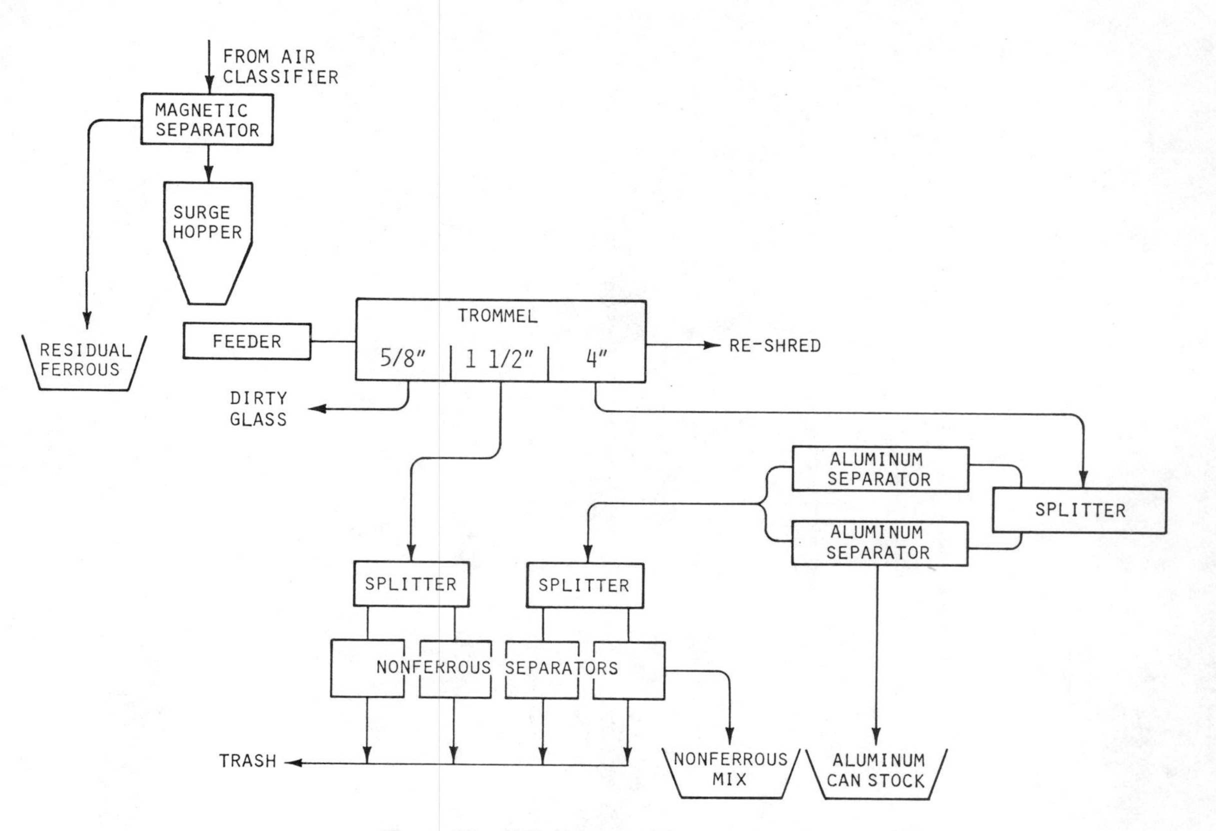

Figure 15. CPU-400 material recovery system.

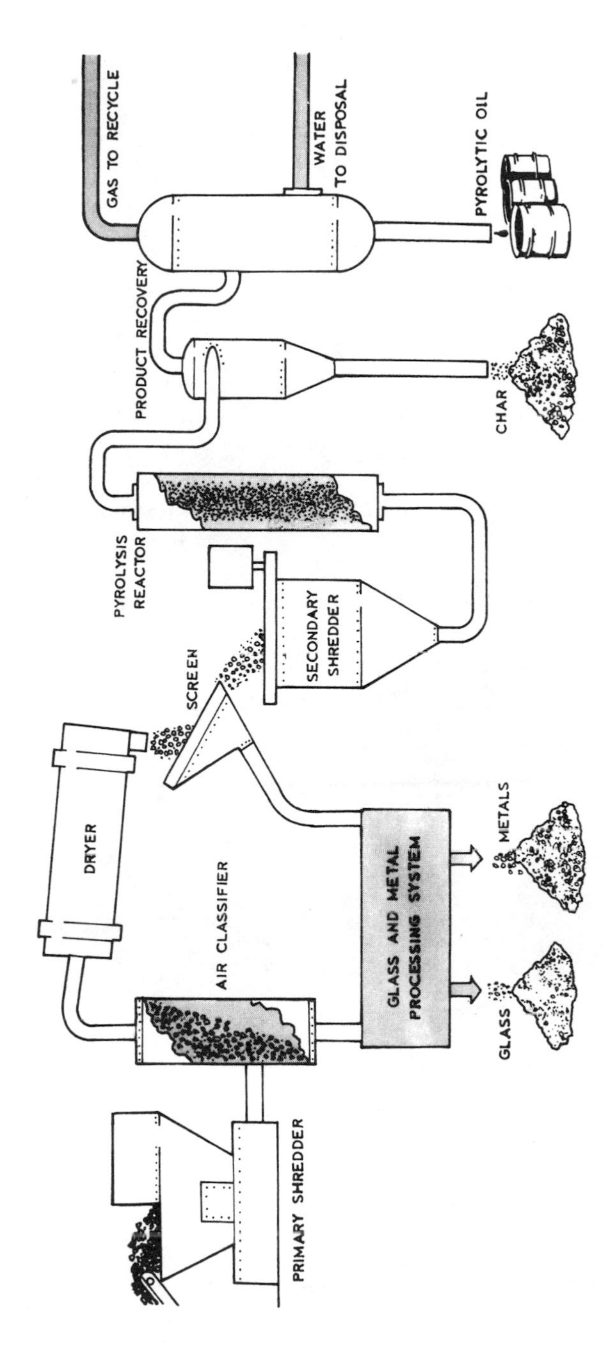

Figure 16. Occidental resource recovery and pyrolysis process.

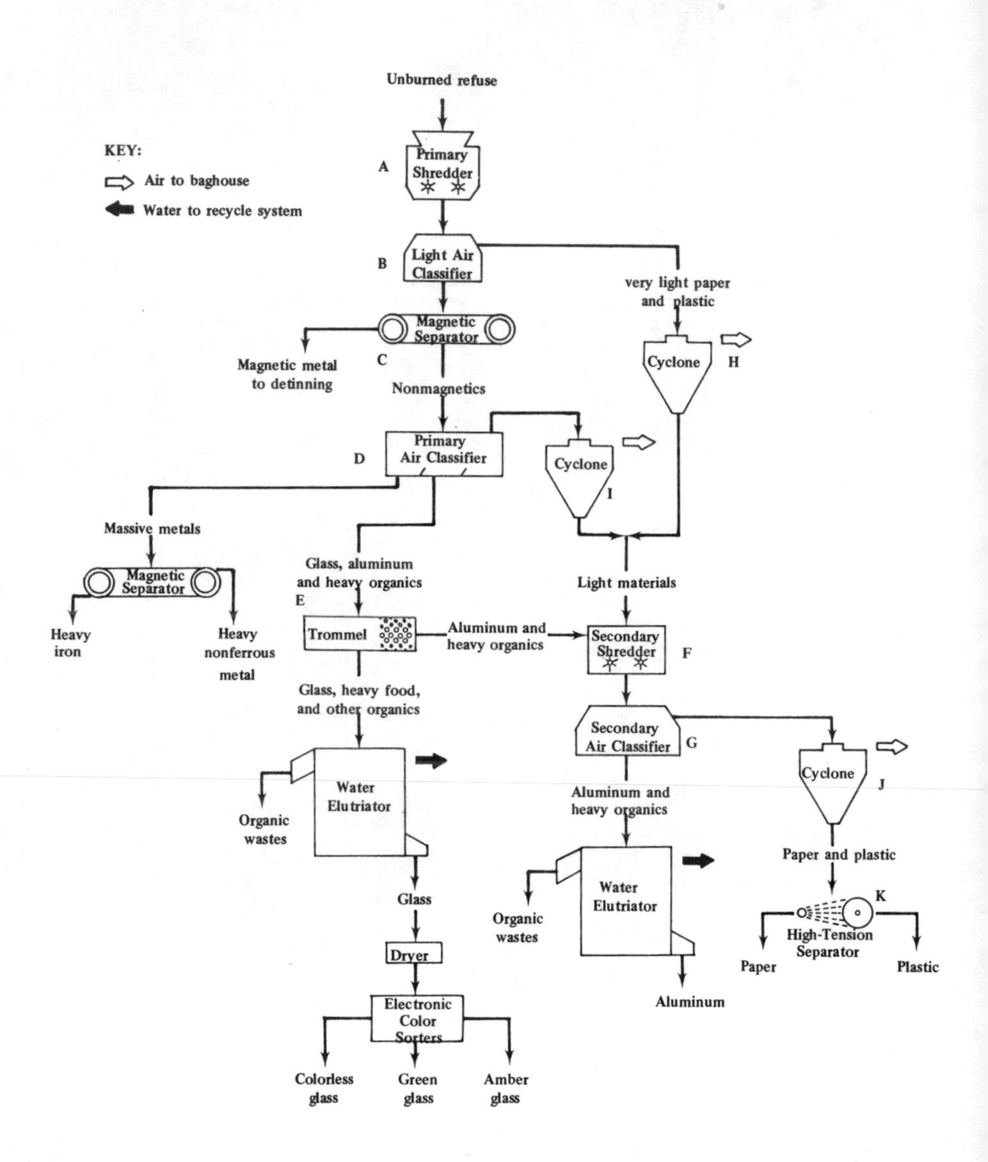

Figure 17. Bureau of Mines raw solid waste separation system.

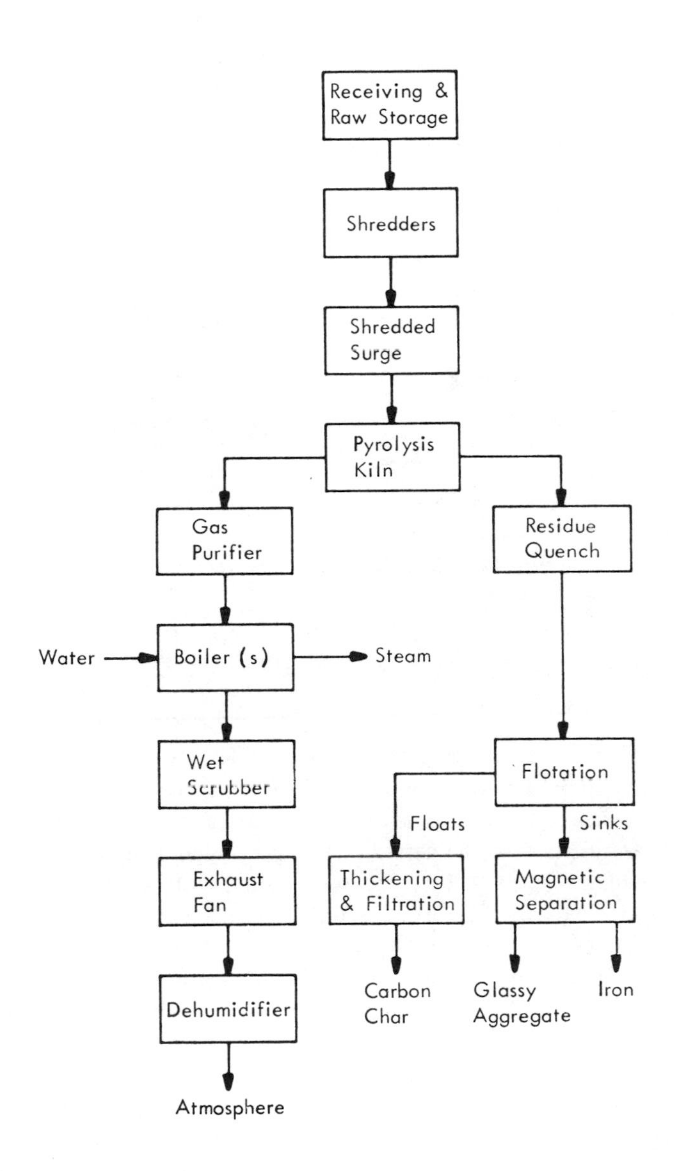

Figure 18. LANDGARD Resource recovery and pyrolysis system.

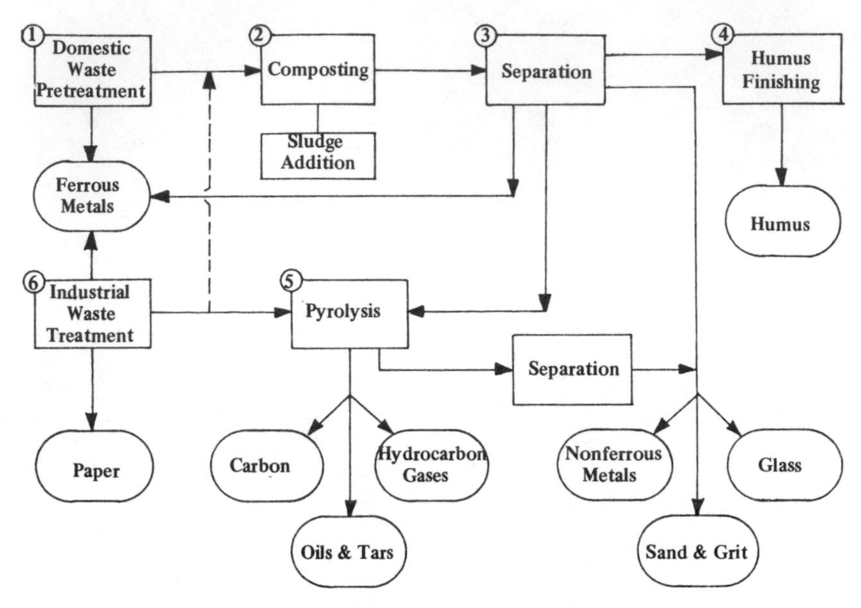

Figure 19. Hercules resource recovery system.

RESOURCE RECOVERY AFTER THERMAL PROCESSING

Each available thermal process produces at least one solid residual material with limited potential value. Typical incinerator residues, shown in Tables 33-35, contain large amounts of glass and ferrous metal. Resource recovery prior to incineration, not now practiced to any significant degree, would obviously reduce the total amount of residue considerably and change its nature. The exact effect would depend upon: whether only ferrous metal was recovered; whether both ferrous metal and glass were recovered; whether other materials such as nonferrous metals were recovered; and the efficiency of these recoveries. The separation and firing of a combustible fraction with coal in a conventional boiler leads to an ash, mixed with an essentially indistinguishable from the coal ash.

The residues from pyrolysis processes were discussed in Chapter 3. These may be high ash chars, with some potential value as low grade fuel, and slag-like materials high in glass content with some potential value as construction materials. Of the commercially-available processes, only in the Landgard process is ferrous metal recovered after pyrolysis, and even there, plans to magnetically separate the solid waste feed will limit such recovery from the residue.

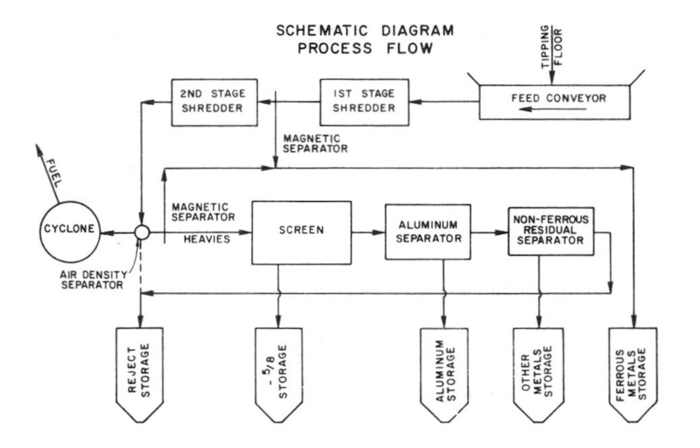

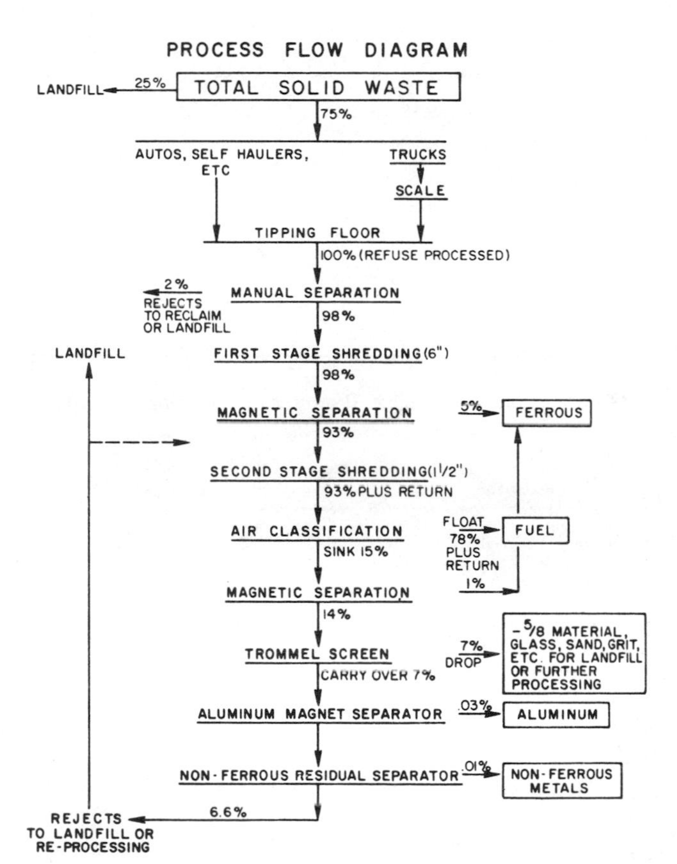

Figure 20. Ames resource recovery system.

Table 33. Composition of Grate-Type Incinerator Residues[2]

Component	Average Percentage (dry weight basis)
Glass	44.1
Tin Cans	17.2
Mill Scale and Small Iron	6.8
Iron Wire	0.7
Massive Iron	3.5
Nonferrous Metals	1.4
Stone and Bricks	1.3
Ceramics	0.9
Unburned Paper and Chemicals	8.3
Partially Burned Organics	0.7
Ash	15.4
Total	100.3

Table 34. Composition of Rotary-Kiln Incinerator Residues[2]

Component	Average Percentage (dry weight basis)
Fines, minus 8-mesh (ash, slag, glass)[a]	35.8
Glass and Slag, plus 8-mesh[b]	21.2
Shredded Tin Cans	19.3
Mill Scale and Small Iron	10.7
Nonmetallics from Shredded Tin Cans	6.5
Charcoal	3.4
Massive Iron	1.9
Iron Wire	0.5
Ceramics	0.2
Handpicked Nonferrous Metals	0.1
Total	99.6

[a]Of the total weight of this fraction, 1.8 percent is recoverable nonferrous metal.
[b]Of the total weight of this fraction, 1.4 percent is recoverable nonferrous metal.

Table 35. Analysis of Incinerator Residue[24] (Dry Basis)

Component	Weight Percent
Wire and Large Iron	3.0
Tin Cans	13.6
Small Ferrous Metal	13.9
Nonferrous Metal	2.8
Glass	49.6
Ash	17.1
Total	100.0

Current Incinerator Residue Salvage

The only significant salvage practiced on incinerator residue is ferrous metal recovery. Even this practice is limited to not more than 10-20 incinerators in the U.S. Ferrous metal, primarily cans, is recovered from residue either magnetically, or by the use of revolving cylindrical screens called trommels. The cans are retained by the trommel, while most of the residue passes through for landfill disposal. Ferrous metal recovered by either method may be washed and shredded in the incineration plant, or by the purchaser.

Most of the recovered ferrous metals must be shipped to the western part of the U.S. for use in copper precipitation. The remaining recovered ferrous metal is either used directly in steelmaking, or first detinned and then used for steelmaking.

Emerging Incinerator Residue Salvage Technology

Ferrous metals are valuable and relatively easy to recover from incinerator residues, but glass and nonferrous metals in the residue are also potentially valuable. For this reason, the U.S. Bureau of Mines has piloted a complete resource recovery system for incinerator residue. A full-size processing plant designed by the Raytheon Company based on Bureau of Mines data, and capable of handling 227 metric tons of incinerator residue per 8 hours, is being built in Lowell, Massachusetts with federal assistance.[12] The processing scheme will be similar to that shown in Figure 21, producing the materials shown in Table 36.

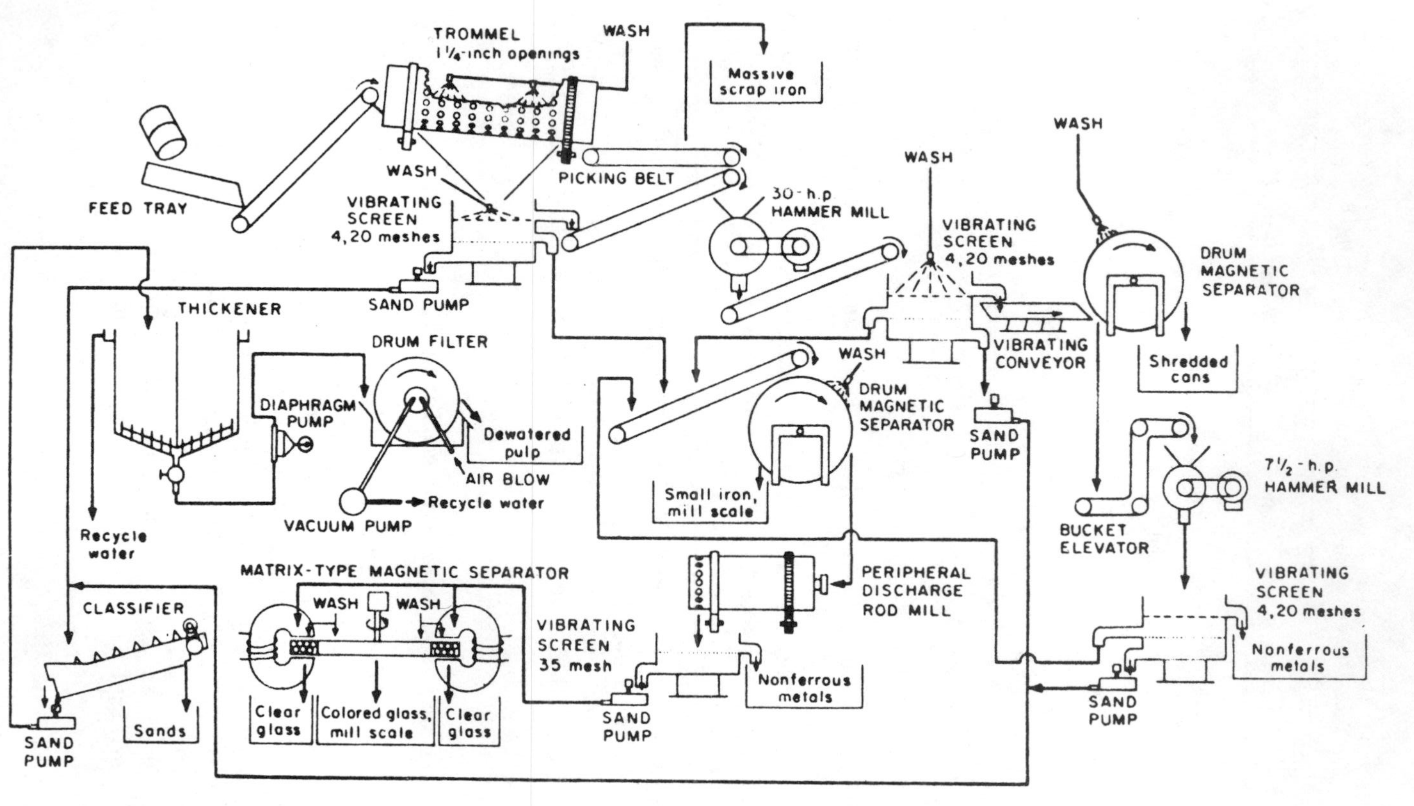

Figure 21. Bureau of Mines incinerator residue recovery system.[31]

Table 36. Expected Products from the Lowell Incinerator Residue Resource Recovery Project

	Weight Percent of the Residue
High Value Products	
Aluminum	1.5
Zinc-Copper	1.0
Ferrous Products	30.5
Colorless Glass	20.8
Medium Value Products	
Mixed Colored Glass	10.2
Slag	14.0
Sand	17.0
Waste Products	
Unburned Organics	1.4
Filter Cake	3.6
	100.0

MARKETS FOR RECOVERED MATERIALS

The ultimate success of any resource recovery system hinges both on the availability of suitable markets, and on the availability of technology and operating skill to produce products meeting the specifications required in the marketplace. The commitment of capital for resource recovery facilities must be preceded by identification, understanding, and securing of markets. A few of the fundamental considerations follow.

Fuel

As previously discussed, the marketing problems which exist for 100% solid waste firing in a steam-producing municipal incinerator are those of being reasonably close to the demand for steam or electrical power, and of matching supply with a varying demand. On the other hand, the preparation of combustible refuse as a fuel, for firing with fossil fuels in existing utility system boilers, introduces normal product marketing problems such as quality, quantity, shipping distance, storage capacity, and competition with conventional fuels.

The experience already accumulated on preparing refuse-derived fuel from solid waste and firing this fuel with coal in an existing boiler lends confidence to such projects. However, as similar projects are undertaken, great care must be observed to avoid problems which could arise due to differences in the physical and chemical properties of the fuel, and due to differences in steam boiler design. For example, chemical properties of the fuel might vary with classifier operation, raising questions about corrosion, slagging, and air pollution control in the boiler. Besides careful choice of processing equipment, the best means for avoiding such problems are careful analysis of raw refuse, prepared fuel, and other materials, and cooperative testing with potential customers.

It is anticipated that carefully prepared fuel from municipal solid waste should eventually approach the value of coal on a net heating value basis, though more complex evaluation procedures have been developed.[26] The calculation of value per ton of fuel is illustrated in Table 37. Although all characteristics of the fuel (such as physical form, sulfur, ash and chlorine content, and ash fusion point) may play a role in determining its price, the heating value and shipping distance are overriding factors, assuming the basic acceptability of the fuel for its intended use.

A full understanding of the intended use is vital in determining its value to the customer. For example, the low sulfur content of prepared refuse fuel is a credit in boilers having difficulty meeting sulfur oxide emission standards, but may be a liability in boilers already burning low-sulfur coal, where low-sulfur levels can lead to insufficient natural sulfur trioxide gas conditioning required for adequate electrostatic precipitator performance. In other applications the ash content may be critical, where the relatively high ash per unit of heating value in fuels derived from solid waste may increase particulate emissions, or overload existing ash handling systems.

Paper Fiber

In some installations the recovery of paper fiber may be justified even at the expense of feed to the thermal processing facilities. For example, paper fiber recovered from an existing resource recovery system was reportedly sold in the $27-$72/MT ($25-$65/ST) range,[27] certainly competitive with fuel use. Therefore, it is advisable to investigate the cost of paper fiber recovery and possible markets to determine whether such recovery is justified.

Ferrous Metals

Ferrous metal, including cans, recovered from solid waste must compete with other sources of iron and steel scrap. Of the three major markets which exist, the western U.S. copper precipitation market, where recovered

Table 37. Calculation of Potential Value for Fuel Prepared from Municipal Solid Waste Based on Lower Heating Value

	Coal A	Coal B	Coal C	Prepared Fuel[a]	
				Range	Average
Composition, Wt %					
Ash	10.70	13.0	5.00	7.6-31.3	16.8
Sulfur S	4.20	0.5	0.92	0.04-0.28	0.10
Hydrogen	4.35	1.9	5.12	–	(3.55)[a]
Carbon	61.52	70.6	77.13	–	(21.12)[a]
Moisture	10.80	11.0	3.50	11.1-66.3	30.3
Nitrogen	1.25	0.8	1.49	–	–
Oxygen	7.18	2.2	6.84	–	(28.13)[a]
	100.00	100.0	100.00		(100.00)
Higher Heating Value					
cal/g	6278	6239	7639	1274-4218	2768
(BTU/lb)	(11,300)	(11,230)	(13,750)	(2293-7593)	(4983)
Lower Heating Value[b]					
cal/g	5986	6074	7349		2403
(BTU/lb)	(10,774)	(10,934)	(13,228)	–	(4326)
Equivalent Values					
$/10^6 Kcal[c]	6.67	6.67	6.67	–	6.67
($/10^6 BTU)	(1.68)	(1.68)	(1.68)	–	(1.68)
$/metric ton	40.00	40.59	49.11	–	16.06
($/short ton)	(36.29)	(36.82)	(44.55)		(14.57)

[a]Data from Table 30. Balance of composition, shown in parentheses, assumed to be cellulose for purposes of this calculation.

[b]By calculation, correcting higher heating value for unavailable heat due to water in flue gas (from moisture and hydrogen combustion) in vapor rather than liquid form. All heating values assumed at 15.6°C (60°F).

[c]Based on lower heating value.

can metal is preferred because of its high surface area per unit of weight, is usually unavailable because of distance and shipping cost. The detinning market has usually been unavailable because of marginal profit and aluminum contamination; and the steelmaking market is undependable because the contaminants in recovered ferrous metal make this a material of last resort.[11] The most favorable counteractants to this picture are the emerging resource recovery technology for separate recovery of other metals and clean ferrous metal recovery, and the emergence of industrial organizations specializing in the cleaning and purification of crude recovered metals for resale to processors. In some cases, compaction of recovered metal to improve handling may be necessary.

To assure stable markets for recovered ferrous metal, cooperative testing and long-term contracts with purchasers are recommended, even at prices well below those which sometimes exist during peak scrap demand. Escalation clauses should be considered based on market conditions and contamination level. A stable supply of reasonably priced scrap may encourage foundries and steelmakers to devise systems with the capability of handling contaminated ferrous scrap. Both the potential for use and the problems have been demonstrated in extensive test work.[28]

Aluminum

Increasing costs for electrical power and for aluminum ores have created considerable incentive for maximizing aluminum recycle. The high price being offered for scrap aluminum has encouraged recovery at the source through volunteer organizations and others.[11] Although it is expected that a considerable amount of aluminum will still find its way into solid waste, its concentration in the waste is very low, on the order of 1 wt % or less. As shown earlier in this chapter, even at low concentration, it is well worth recovering because of the high price. As with ferrous metals, contamination is a very important consideration, increasing reclaiming costs, and careful attention must be given to this problem.

An interesting possibility exists for installing aluminum scrap electrical melting furnaces to produce aluminum ingot for sale to fabricators. In thermal processing facilities where electrical power is produced, this possibility is even more attractive. Such an operation could handle both aluminum recovered from solid waste and aluminum separated at the source, for example from volunteer organizations.

Other Nonferrous Metals

As shown earlier in this chapter, other nonferrous metals such as copper and zinc can be recovered from solid waste as mixed metallics, or possibly as separate metals by further processing. Many nonferrous metals are in short supply worldwide, providing considerable incentive for improving recycling methods.

Glass Cullet

A good market exists for clean color-sorted glass cullet in many parts of the U.S. These materials are used in existing glass melting furnaces along with the raw materials normally used in glass making. The development of adequate methods to recover clean glass sorted into flint (colorless), amber, and green colors is the critical problem in this market, but present technology appears to be expensive. If good quality glass can be produced consistently, it should find ready market acceptance in glass melting furnaces, because of savings in energy consumption and advantages in air pollution control as compared to the use of raw materials for glassmaking. The potential for unsorted glass is much less favorable.

Plastics

Although no market now exists for plastics recovered from solid wastes, it is possible that as technology develops for recovering plastics, markets could become available either through existing reclaiming operations, or by the development of new products which can use the reclaimed plastics. Certainly the high price paid for clean segregated plastic scrap provides considerable incentive for improved technology.[11] Unfortunately, the very wide variety of plastic materials and the low concentration present in solid waste preclude any simple answer to the problems of separation and salvaging. At the moment it appears that, at least in thermal processing facilities, the best outlet for plastics is in the thermal process itself, either recovering heat from combustion or fuels by pyrolysis.

Flyash

The principal sources of flyash in thermal processing facilities will be from air pollution control equipment in incinerators, and in boilers where fossil fuels and prepared refuse are fired together. These ashes may be available wet or dry depending upon the particular forms of air pollution control and of ash handling. Some important uses of flyash from coal-fired steam boilers have been developed, primarily in construction materials, but these uses consume only a minor part of the total available ash.[29]

Flyash from incinerators will compete in the same market, either successfully or unsuccessfully, depending upon the particular location and aggressiveness of the facility management. Fortunately, sterile flyash is not an objectionable fill material and can be disposed of where fill is desired, in special landfill sites, or in sites where a variety of wastes are accepted.

REFERENCES

1. Resource Recovery–The State of Technology. Prepared for the Council on Environmental Quality by Midwest Research Institute. National Technical Information Service. Springfield, Va. PB-214 149. February 1973. 67 pages.
2. Drobny, N. L. *et al.* Recovery and Utilization of Municipal Solid Waste. Report No. SW-10c. U.S. Environmental Protection Agency. U.S. Government Printing Office. Washington, D. C. 1971.
3. Shredders. . . Processing Our Solid Waste. The NCCR Bulletin. National Center for Resource Recovery (Washington, D.C.) 111(1): 12-18, Winter 1973.
4. Dale, J. C. Recovery of Aluminum from Solid Waste. Resource Recovery. Jan/Feb/Mar/1974. Pages 10-15.
5. Cheremisinoff, P. N. Air Classification of Solid Wastes. Pollution Engineering. December 1974. Pages 36-37.
6. Boettcher, R. A. Air Classification of Solid Wastes. U.S. Environmental Protection Agency. SW-30c. U.S. Government Printing Office. Washington, D. C. 1972. 73 pages.
7. Campbell, J. A. Electromagnetic Separation of Aluminum and Nonferrous Metals. Combustion Power Company, Inc. (Presented at 103rd. American Institute of Mechanical Engineers Meeting. Dallas, Texas February 24-28, 1974.
8. AL MAG 40-Aluminum Magnet Separator Systems. Combustion Power Company, Inc. Menlo Park, Calif.
9. Solid Waste Air Classifier–Model 50. Combustion Power Company, Inc. Menlo Park, Calif.
10. Neff, N. T. Solid Waste and Fiber Recovery Demonstration Plant for the City of Franklin, Ohio. Prepared for U.S. Environmental Protection Agency by A. M. Kinney, Inc. PB-218 646. National Technical Information Service. Springfield, Va. 1972. 83 pages.
11. Darnay, A. and W. E. Franklin. Salvage Markets for Materials in Solid Wastes. Contractor-Midwest Research Institute. Kansas City, Missouri. U.S. Environmental Protection Agency. SW-29c. U.S. Government Printing Office. Washington, D. C. 1972. 187 pages.
12. Office of Solid Waste Management Programs. Second Report to Congress–Resource Recovery and Source Reduction. U.S. Environmental Protection Agency. SW-122. U.S. Government Printing Office. Washington, D. C. 1974. 112 pages.
13. Sullivan, P. M. *et al.* Resource Recovery from Raw Urban Refuse. Bureau of Mines. RI 7760. U.S. Government Printing Office. Washington, D. C. 1973. 28 pages.

14. Herbert, W. and W. A. Flower. Glass and Aluminum Recovery in Recycle Operations. Public Works. August 1971.
15. DiGravio, V. P. Materials Handling and Shredding Systems for Size Reduction of Solid Waste Constituents. Metcalf & Eddy, Inc., Boston, Mass. (Presented at American Society of Mechanical Engineers Design Committee Meeting, January 20, 1971). 9 pages.
16. CEA's Brockton Plant Now Producing Fuel. Resource Recovery. Jan/Feb/Mar/1974. Page 30.
17. Winners Named for Construction and Operation of $80,000,000 Resource Recovery Plants Connecticut. Resource Recovery. Apr/May/June/1974. Pages 8-9.
18. Small Pellets Made From the Nation's Wastes Could Help Supplemental Fuel Supplies. The American City. March 1974. Page 137.
19. Grubbs, M. R. and K. H. Ivey. Recovering Plastics from Urban Refuse by Electrodynamic Techniques. Technical Progress Report 63. Bureau of Mines. PB-214 267. National Technical Information Service. Springfield, Va. December 1972. 6 pages.
20. Can a Smaller City Find Happiness With Resource Recovery. Resource Recovery. Nov/Dec/1974. Pages 8-12.
21. American Can Will Take Over Disposal of Milwaukee's Solid Waste. Chemical Week. January 22, 1975. Page 35.
22. Liabilities Into Assets. Environmental Science and Technology. 8(3):210-211. March 1974.
23. Trash-Can Contents Turned Into Fuel. The American City. March 1974. Page 137.
24. Henn, J. J. and F. A. Peters. Cost Evaluation of a Metal and Mineral Recovery Process for Treating Municipal Incinerator Residues. IC 8533, Bureau of Mines. U.S. Government Printing Office. Washington, D. C. 1971. 41 pages.
25. Klumb, D. L. Solid Waste Prototype for Recovery of Utility Fuel and Other Resources. Union Electric Company. Paper APCA 74-94. (Presented at Air Pollution Control Association 67th Annual Meeting. Denver, Col. June 9-13, 1974). 16 pages.
26. Eggen, A. C. and R. Kraatz. Relative Value of Fuels Derived from Solid Wastes. Proceedings, 1974 National Incinerator Conference. Miami, Fla. May 12-15, 1974. American Society of Mechanical Engineers. pages 19-32.
27. Colonna, R. A. and C. McLaren. Decision-Makers Guide in Solid Waste Management. SW-127. U.S. Environmental Protection Agency. U.S. Government Printing Office. Washington, D. C. 1974. 157 pages.
28. Ostrowski, E. J. Recycling of Ferrous Scrap from Incinerator Residue in Iron and Steel Making. Proceedings, 1972 National Incinerator Conference. New York, N. Y. June 4-7, 1972. American Society of Mechanical Engineers. Pages 87-96.
29. Capp, J. P. and J. D. Spencer. Fly Ash Utilization—A Summary of Applications and Technology. IC 8483. Bureau of Mines. U.S. Government Printing Office. Washington, D. C. 1970. 72 pages.
30. Sutterfield, G. W. *et al.* From Solid Waste to Energy. City of St. Louis *et al.* (Presented at the U.S. Conference of Mayors. Solid Waste Seminar. Boston, Mass. October 4, 1973). 13 pages.
31. EPA Supports Incinerator Resource Recovery. Reuse/Recycling 2 (No. 8):2. Technomic Publishing Co. Westport, Conn. Dec. 1972.

CHAPTER 5

THERMAL PROCESSING COSTS

As for any other process plant, thermal processing costs consist of those required to:

1. Acquire the necessary facilities (capital costs),
2. Own the plant (amortization and interest on capital costs), and
3. Operate the facilities (labor, utilities, supplies, maintenance, overheads).

With the surgent interest in energy and resource recovery, another major cost dimension has been added. The recovery of "by-products" from solid waste disposal normally increases costs for all three of the above, but sales revenues from by-products can also provide credits against operating costs.

Other financing arrangements which are sometimes advantageous and feasible, such as private ownership and operation, will not be dealt with in this publication. Provisions for working capital and depreciation may also be considered, depending on cost management procedures used by a particular municipality.

Accurate pre-construction and post-construction cost data are essential for planning; for funding decisions; for municipal budgets and accountability; to aid in decision-making for plant modifications; to set solid waste disposal prices for private or public parties not involved in ownership; and to negotiate by-product prices, for example, for steam from energy recovery incinerators, fuels from pyrolysis facilities, and for glass, metals, and other resources recovered from a variety of thermal processing facilities. A useful accounting procedure for incineration operations is available.[1]

It cannot be emphasized enough that the data presented herein are meant for orientation and illustration only. Even for planning purposes, study cost estimates prepared by qualified design and construction firms for specific facilities must be obtained. The recent rapid rate of inflation, changing technology, increasingly stringent environmental and worker health and safety requirements, and other such factors, may make generalized cost data almost immediately obsolete.

CAPITAL COSTS FOR REFRACTORY INCINERATORS

Capital costs necessary to acquire conventional incinerator facilities can be broken down as follows:

- pre-operating expenses such as legal, financial, and consulting fees, and interest on construction loans
- land acquisition and site preparation, including fences, roads and parking lots
- engineering, project management, construction expenses, and contractors' fees
- buildings and foundations
- refuse weighing, handling, preparation and storage systems
- furnaces and appurtenances
- fans, pumps and motors
- residue removal systems
- air pollution control systems, including stack(s)
- water pollution control systems
- utility generation and distribution
- instrumentation, controls, and laboratory
- piping and duct work
- locker rooms, offices and sanitary facilities
- start-up costs, including acceptance tests

Each of these cost elements is significant and requires careful consideration as discussed elsewhere in this publication. When budgeting capital costs, a contingency should also be included. The contingency is a judgment factor which may range from as little as 5%, based on estimates for completely designed systems, to as much as 20-30% for systems in the planning stage which involve incompletely developed technology.

A 1969 study[2] developed capital cost ranges for batch and continuous feed incinerators. As shown in Figures 22 and 23, these costs are grossly broken down into furnace, building and electrostatic precipitator costs, including most of the above items but not pre-operating expenses, land acquisition and site preparation, or project management.

The costs shown in Figures 22 and 23 can be corrected to the year in question using Marshall & Swift (M&S) Indexes[3] (annual average) which follow:

1968	273.1	1972	332.0
1969	285.0	1973	344.1
1970	303.3	1974	398.4
1971	321.3	1975	445 (preliminary)

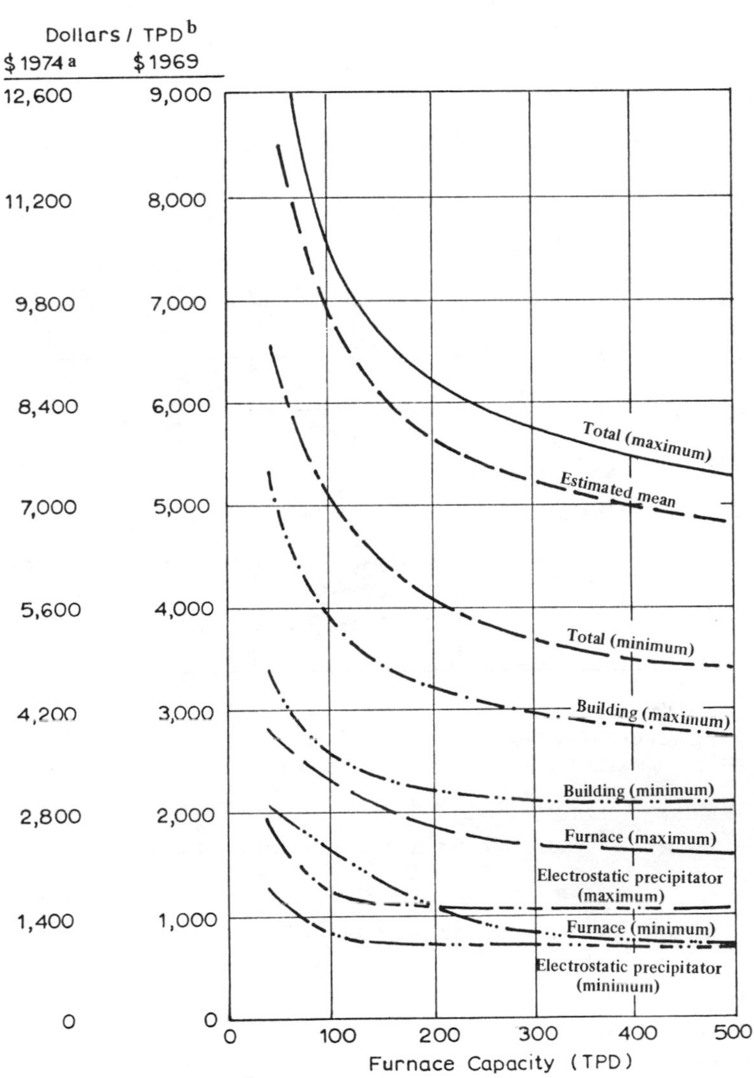

[a]M&S factor 1969→1974 = 1.40[3]
[b]TPD = short tons (2000 lbs) per day. TPD x 0.0378 = metric tons per hour

Figure 22. Capital costs for refractory batch fed incinerators.[2]

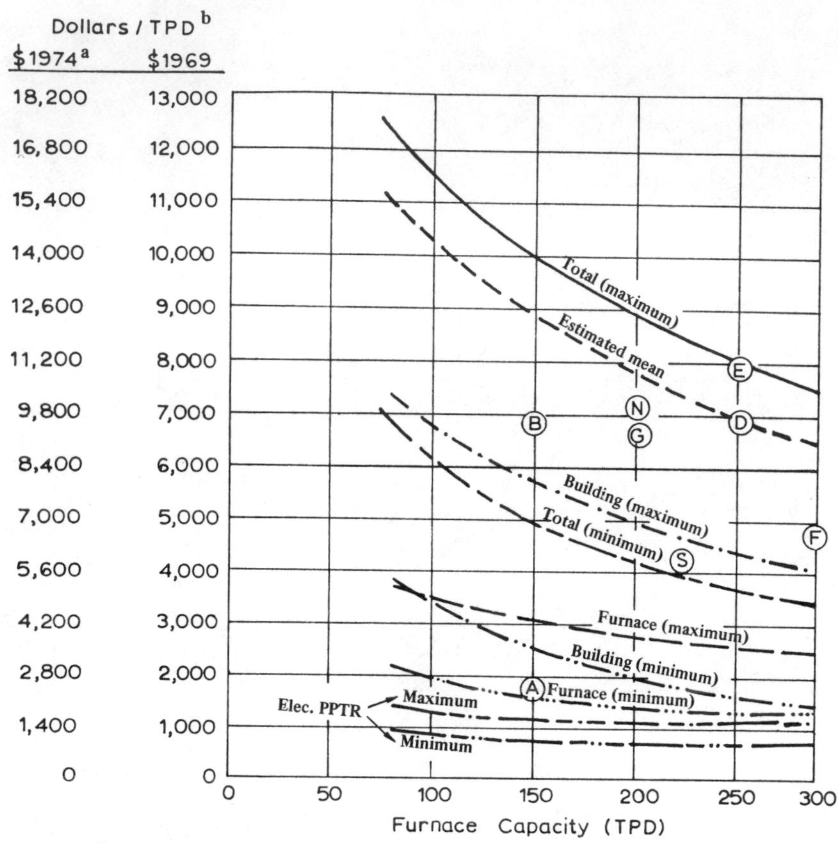

[a]M&S factor 1969→1974 = 1.40[3]
[b]TPD = short tons (2000 lbs) per day, TPD x 0.0378 = metric tons per hour

Figure 23. Capital costs for refractory continuous fed incinerators.[2]

Capital costs for eight incinerators built before 1968, obtained from the literature[4] and other sources, corrected by use of M&S Indexes and plotted on Figure 23, agree well with the curves for total investment. However, the wide range of possible costs,[2,4] recent stringent standards, as well as previously noted uncertainties, point up the need for careful cost estimation, and cost-benefit analyses for discretionary design factors.

CAPITAL COSTS FOR STEAM-GENERATING INCINERATORS

Steam generation, using municipal solid waste as fuel, is receiving an unprecedented surge of interest in the U.S. with no fewer than 20 cities planning, building or operating facilities to produce steam in specially-designed incinerators, or to burn prepared refuse in fossil fuel-fired boilers.

As compared to refractory furnaces, costs for the following items must be added for steam-generating incinerators:

- waterwall furnaces (replacing refractory furnaces)
- waste heat boilers, high-pressure steam piping, soot blowers, and
- other appurtenances
- turbine drives for in-plant steam use where feasible
- boiler feedwater and condensate treatment systems
- excess steam condensers
- complete auxiliary boilers and/or burners
- boiler controls and instrumentation

Although the use of waterwall steam-generating boilers reduces gas flow and therefore the cost of the gas handling equipment [*e.g.*, fans, air pollution control equipment, stack(s)], the above noted cost elements more than offset these reductions, increasing overall capital cost. With present day fuel prices, the increased capital cost can be justified by steam sales where a market exists.

Referring to Figure 24, a reproduction of the 1969 study data[2] on steam-generating incinerators, the estimated 1974 cost can be compared with a refractory continuous fed incinerator from Figure 23, as shown in Table 38.

In Table 39, actual cost data for three steam-generating incinerators were adjusted to 1974 and compared to the 1969 cost curves by plotting the data on Figure 24. In this case, the capital costs for these recent incinerators exceeded values predicted by the curves. This is most likely due to more stringent design standards, especially with regard to air pollution control equipment.

When the municipality is responsible for steam distribution, an additional major capital cost may be incurred, depending on distance and steam pressure. If the steam is converted to another form of energy, such as chilled water or electric power, major capital costs are also required for energy conversion facilities and distribution. An example of such costs is provided in Table 40.

The major capital costs for a steam-generating incinerator are for furnaces, steam boiler equipment, residue handling, air handling, and air pollution control. Together these comprised over 70% of the total cost for a recent installation, as shown in Table 41.

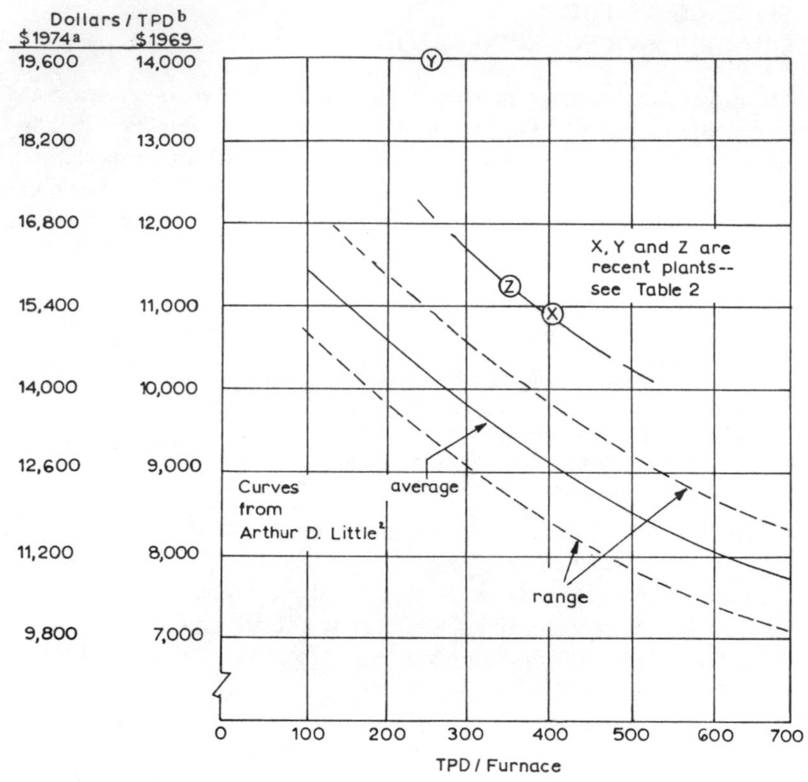

Figure 24. Capital costs for steam generating incinerators (excluding steam distribution).

Table 38. Mean 1974 Cost Per Ton of Capacity at 11.3 MT/hr (300 TPD) Capacity[a]

Refractory Incinerator (from Figure 23, maximum)	\$280,000 per MT/hr (\$10,600/TPD)
Steam-Generating Incinerator (from Figure 24, high end of range)	\$392,000 per MT/hr (\$14,800/TPD)

[a]1 MT/hr = 1 metric ton per hour = 2205 lb/hr.
1 TPD = 1 short ton per day = 2000 lb/day.
Excluding energy distribution or conversion system costs.

Table 39. Actual Costs for Three Steam-Generating Incinerators[a]

Site	X	Y	Z
Capacity, TPD	4 x 400	4 x 250	2 x 360
Actual Cost, $MM	17.4	16.8	9.8
Year	1969	1973	1973
M&S Cost Factor	1.40	1.16	1.16
1974 Cost, $MM	24.4	19.5	11.4
1974 Cost, $/TPD	15,300	19,500	15,800
$ per MT/hr	403,000	516,000	419,000

[a]TPD = ST/day

Table 40. Actual Costs of Energy Distribution and Conversion Systems[a]

Site	Y	Z
Year	1973	1973
Refuse Capacity, TPD	4 x 250	2 x 360
Distribution System	steam	steam, chilled water
Distribution System Cost	$2.2 MM	$4.0 MM
Energy Conversion System	none	steam → chilled water
Energy Conversion System Cost	–	$3.0 MM

[a]TPD = ST/day

Table 41. Capital Cost Breakdown For a Steam-Generating Incinerator[a]

	%
Furnaces, boilers, precipitators, ID fans, ash conveyors, ash crane	70.7
Building, foundations and concrete works	7.4
Building, steel structure	2.2
Building, general construction	4.7
Refuse cranes	2.5
Chimney and flyash silo	1.4
Conveyor system for flyash	1.1
Pumps and steam turbine	0.6
Emergency steam condenser	1.5
Electronic weighing scales	0.5
Water treatment plant	1.0
Central control panel and instrumentation	0.5
Other plant utility equipment and systems (fuel oil, air, steam, electrical)	1.5
Access ramps, water and sewer	2.4
Landscaping and site works	1.2
Temporary services during construction	0.8

[a]Derived from reference 5. Excludes land cost, engineering, contingencies, and steam transmission.

CAPITAL COSTS FOR OTHER TYPES OF THERMAL PROCESSING UNITS

Since pyrolysis facilities are only now approaching commercialization, no historical cost data are available. It is reported that the Baltimore 37.8 metric ton per hour (1000 short tons/day) facility, using the Monsanto Landgard process, has cost 16 million dollars.

The prototype fossil fuel/prepared refuse combustion (co-combustion) project was reported to have cost 3.5 million dollars, including 2.95 million for a 12.3 metric ton per hour (325 short ton/day) refuse preparation system and 0.55 million for receiving and handling facilities at the existing power plant.

Recent cost projections have been made for 37.8 MT/hr (1000 ST/day) co-combustion and PUROX pyrolysis systems[6]:

	$ per MT/hr	$ per ST/day
Co-combustion	251,000-291,000	9,500-11,000
PUROX	627,000-693,000	23,700-26,200

The co-combustion process produces a refuse-derived fuel and salvage materials, while the PUROX process, currently being tested in a large pilot plant, produces a clean fuel gas and fused frit, with the option of materials recovery instead of frit.

OPERATING COSTS

The operating cost for thermal processing facilities, as used in this publication, is the expense involved in keeping the facility running to dispose of solid waste and to recover products of value, when the latter is an integral part of the operation. Operating costs are usually broken down into direct (or variable) costs and indirect (or fixed costs), although some costs are semi-variable.

Direct costs, such as utilities and residue disposal, tend to be proportional to solid waste throughput. Indirect costs, such as insurance and facility protection, tend to be independent of throughput. When projecting or otherwise analyzing costs, it is necessary to carefully determine which costs for the specific facility in question are indirect or direct. For example, although operating labor is normally considered a direct cost, operation of a thermal processing facility with municipal employees may require the maintenance of a fixed size labor force over a long period of time, regardless of throughput. Thus, normal wages for such a labor force become an indirect cost, independent of throughput.

Direct Labor and Labor Overhead Costs

As shown in Table 42, labor and labor overheads comprise the largest single operating cost. Unit labor costs ($/ton) are obviously a function of wage rates and the number of personnel required, but also depend upon actual *vs.* design throughput where the total number of personnel is rather inflexible. Wage rates usually are fixed by prevailing scales paid to comparable municipal employees. However, higher wage scales competitive with local industry, where different than municipal scales, may attract personnel of greater experience, training, and responsibility.

Table 42. Example of Operating Cost Calculation for Incinerator Operating at Design Capacity

Basis:
1. 16 MT/hr average output, 24 hrs/day, 365 days/yr
2. 50 KWH/MT electric power @ 4¢/KWH
3. 5 MT/MT cooling water @ 2.5¢/MT
4. Direct labor, 50 men @ $13,000/yr average wage
5. Labor overheads at 30% of direct labor
6. Contract maintenance and materials, and supplies, @ $150,000/yr
7. Indirect costs at 40% of direct labor

	Operating Costs,* $/MT
Direct	
Labor	$ 4.64
Labor Overheads	1.39
Utilities	2.13
Maintenance	1.07
	9.23
Indirect	1.86
	$ 11.09

*Excludes ownership costs.

Since increasing facility size does not proportionately increase personnel requirements, large facilities show smaller unit labor costs than small ones. For example, although three incinerator operators per shift may be required for three 400 ton per day incinerator trains, four similar size trains may also require only three operators, a reduction of 25% in unit cost for capacity operation.

Labor requirements are also determined by the degree of instrumentation, automatic control, and other labor-saving devices. Similarly, added capital investment for spares, high quality equipment components, and other methods for improving reliability can also reduce unit labor costs, but each such added investment should be carefully justified, using actual operating data where possible.

Overheads which are directly related to labor requirements, for example, payroll taxes and benefits such as retirement and health plans, vacation, sick leave, etc., are termed labor overheads. These vary with location but are usually identical to those used for other munitipal employees, except where special benefits such as safety shoes, safety glasses and hardhats are supplied by the incinerator management.

When calculating labor costs, provision must be made for overtime pay, shift differentials, and other costs associated with continuously manning an operating facility.

Utility and Direct Supply Costs

Utility and supply requirements and costs for individual incinerators may differ markedly. Some of the factors which affect these costs are shown in Table 43.

Table 43. Major Factors Affecting Incinerator Utility Costs

Factor	Comments
Local Utility Price Structure	Prices for purchased utilities such as electric power and water vary greatly between localities, due to differences in availability of fuels for power generation, and of natural water supplies.
Excess Air Requirements	Power requirements increase as excess air increases, both for forced draft fans and for induced draft fans. Steam-generating incinerators with waterwalls use less excess air.
Type of Air Pollution Control	High energy scrubbers for particulate control have high draft requirements, provided by fans which consume much more power than is necessary for the combined requirements of electrostatic precipitators and fans for that type of system.
Internal Generation of Steam	A major portion of required power can be supplied by internally-generated steam. Chemicals are required for boiler feedwater treating.
Residue and Wastewater Systems	Makeup-water requirements can be minimized by reusing wastewater from scrubbers and spray coolers to handle solid residues, and by effective wastewater treatment systems which allow maximum water recycle.
Solid Waste Composition	Increased waste moisture in refractory incinerators can decrease excess air required for cooling, but too high a moisture content can necessitate auxiliary fuel burning, especially in steam-generating incinerators. Acidic precursors in waste, such as polyvinylchloride plastics, affect requirements for neutralizing chemicals in wastewaters.

The major utilities normally required in incinerators are electric power for motors and lighting, fuel for space heating and auxiliary steam production, and water for cooling, quenching, drinking, steam generation, and sanitary facilities. More than one type of water may be used, for example, municipal water for drinking and river water or sea water for residue quenching. Direct supplies used in incinerators may include chemicals for water treating, charts and other supplies for instruments, janitorial supplies, deodorants, personal safety equipment, uniforms, and a myriad of other small items. Usually excluded from this category are materials used in repair and maintenance.

When prices do not vary substantially with the quantity used, utility and supply costs may be considered direct, these costs being dependent on throughput. The estimate of utility costs for new incinerators should be based on local projected rates and on sound engineering estimates of the quantities required. Supply costs are normally much less significant than utility costs.

Maintenance Costs

The two major components of maintenance costs are materials and labor used for repairs and routine maintenance, although where contract maintenance is practiced this may be considered as a separate category. The maintenance of instrumentation, cranes, weighing scales, and other complex equipment by outside contractors should be considered because it is difficult to find all necessary skills in the relatively small maintenance crews available at most incinerators. Centralized maintenance for all public works is used in some municipalities.

Maintenance costs vary greatly with adequacy of the incinerator design, age of the facility, quality of equipment, skill of the operators, and nature of the solid waste. For example, replacement of refractory, a major cost in most incinerators, is dependent on all these variables, with minimal costs incurred where high quality refractory is used and where automatic temperature control is reliable.

Scheduled inspection and maintenance programs help to hold down the extraordinary cost sometimes associated with unexpected equipment breakdown and sudden loss of incineration capacity. Total maintenance costs can be expected to lie in the range of 1-5% of the original investment per year, although costs may be expected to increase with the age of the facility faster than escalation, and vary greatly from year to year.

Other Overhead Costs

Overhead costs, other than labor overheads previously discussed, are those costs that are necessary but not easily connected directly with the operation of the facility. They are, in fact, indirect or fixed costs incurred whether or not the incinerator is operating. These may include management, accounting, engineering, secretarial and clerical personnel and costs, insurance, laboratory expenses, training, travel, and other costs. The distinction between overheads and direct costs is not always clear, but should be made as consistently as possible. Depending on accounting methods and the distinctions made, overhead costs may run, for example, from 30% to 60% of direct labor.

OWNERSHIP COSTS FOR INCINERATORS

Ownership costs may be described as those costs which accrue whether the facility operates or not, temporarily or permanently. The ownership costs for municipal thermal processing facilities are interest payments on borrowed capital and the return of that capital. If money is actually set aside periodically for the purpose of returning borrowed capital at a future date, for example as prescribed in bonds issued to lenders, the process is called amortization. If the original capital used to build a facility comes from general tax revenues, a yearly depreciation expense should be charged, which allows for the decrease in the value of the facility due to wear and tear and obsolescence, recognizing the need for future capital to replace the existing facility.

Where bond terms are such that interest is paid periodically and an actual amortization sinking fund is established to repay the loan on a certain future date, annual ownership costs are the annual interest cost plus the annual amortization payment to the sinking fund, less interest earned by the sinking fund. Even where no actual sinking fund exists, costs may be calculated in this way. In any case, ownership costs should be calculated in such a way as to be consistent with the terms of securities sold to raise the necessary capital.

When depreciation expense instead of amortization is used to analyze thermal processing ownership costs, the normal practice is to uniformly distribute the total capital cost by annual charges over a period of about 20-30 years, depending on the predicted life of the facility, or to depreciate plant components over periods which range from about 4-30 years depending on expected life. More complex depreciation approaches which recognize greater depreciation in early years (accelated depreciation) are often used in industry, especially for tax advantages, but find little application in thermal processing facilities.

Interest charges will be determined by free market rates and the credit rating of the municipality when bonds are sold. Since the interest on municipal bonds is received tax-exempt by the owner of the bonds, the rates are significantly lower than comparably rated industrial bonds. A typical ownership cost calculation is provided in Table 44. Because of the great significance of ownership costs, more than the usual care should be taken in calculating these, and in being explicit about the calculation basis.

Table 44. Typical Ownership Cost Calculation

Basis:

1. Capital cost (per unit of capacity) = Capital borrowed = \$397,000 per metric ton/hr (\$15,000/TPD)
2. Bond interest rate = 7%/yr
3. Repayment of bond after 30 years
4. 30-year life for incinerator (no salvage value)
5. Sinking fund interest = 5%/yr
6. Operation at 100% of capacity

	Ownership Cost \$/metric ton
Yearly Interest @ 7%/yr	\$ 3.17
Uniform Annual Amortization Payment[a]	0.68
Total Ownership Cost	\$ 3.85

[a]If this payment is made annually for 30 years to a sinking fund earning 5% interest compounded annually, the sinking fund balance at the end of the 30 years will be equal to the original capital borrowed.

Since ownership costs are obviously directly related to the size of the capital investment, increased investment for greater reliability or lower labor, utility or maintenance costs will correspondingly increase the cost of ownership. Therefore, careful cost-benefit analysis is required to determine the optimum investment/operating cost relationship for the specific project being contemplated.

Underutilization dramatically increases the magnitude of indirect and unit ownership costs (\$/ton), since these costs go on even if the facility never operates. For example, Table 45 shows the increase in total unit costs for an existing incinerator operating on different schedules and throughputs. Unit costs of \$14.94/MT for a seven-day, three-shift operation

Table 45. Economic Effect of Underutilization of Incinerator Facilities

Operating Schedule	7 day - 3 shift	5 day - 2 shift	5 day - 1 shift
Operating Rate, % of Capacity	100	47.6	23.8
Design Capacity, MT/hr (ST/day)		16 (425)	
Operating Rate, MT/yr	140,000	66,640	33,320
Operating Costs, $/MT			
Direct	9.23	9.23	9.23
Indirect	1.86	3.91	7.82
Ownership	3.85	8.09	16.18
	$14.94	$21.23	$33.23

increase to $33.23/MT for a five-day, one-shift operation, due to the effect of fixed costs. Obviously, oversizing an incinerator, or underutilization, can result in an extraordinary cost for refuse disposal.

ECONOMICS OF ENERGY AND RESOURCE RECOVERY

Energy may be recovered from thermal processing systems in the form of steam, fuels or electrical power, as shown in Tables 46 and 47. Part of the energy recovered may be used internally, but most is available for export. Other resources, such as glass, ferrous scrap, aluminum and other metals, can be recovered prior to or after thermal processing. Combustible resources such as paper fiber may also be recovered, but are not normally done so as a part of a thermal processing system.

The value of energy and recovered resources can have a major impact on net thermal processing costs, and can theoretically even pay for the entire cost of thermal processing. The following discussion is designed to show the potential promise for energy and resource recovery, but the management of thermal processing facilities must overcome the institutional, technical and marketing problems inherent in realizing this promise.

Table 47 clearly shows the potential for recovery and the effect of price structure on this potential.

The recovery of glass and metals is of definite interest as a method of offsetting thermal processing costs, but it is the recovery of energy in an environment of rising energy prices that shows really major potential for beneficial use of municipal solid waste.

Table 46. Energy Recovery from Municipal Solid Waste Thermal Processes

Type of Thermal Processing	Form of Energy Recovery[a] (based on refuse as delivered)
Refractory Incinerators with or without Waste Heat Boilers	0-1.5 tons steam/ton refuse (or electric power generated from steam)
Modern Waterwall Incinerators	1.5-4 tons steam/ton refuse (or electric power generated from steam)
Combined Fossil Fuel/Refuse Combustion Boilers	1.5-4 tons steam/ton refuse (or electric power generated from steam, *e.g.*, 500-800 KWH per metric ton of refuse)
Pyrolysis Plants	Gaseous, liquid or solid fuels (or steam or power generated from fuels)
High Pressure Fluidized Bed (under development)	400-500 KWH electrical power/metric ton of refuse

[a]Quantitative values for recovered energy derived from reference 7 and other sources.

However, it is insufficient simply to know the value of resource and energy recovery; the cost of such recovery must also be thoroughly evaluated. As shown in Table 48, the cost of separation steps required prior to thermal processing for resource recovery are substantial.

The overall cost of recovery must be estimated with great care for each facility under consideration. However, for purposes of orientation only, a hypothetical study is provided in Table 49.

It is obvious from the information in Table 49 that the critical aspect of an economically attractive project for thermal processing with resource recovery is locating and assuring markets for the recovered energy and materials. The size of the project is also a critical factor.[6]

One recent set of cost estimates (1974) made for projects producing refuse-derived-fuel, pyrolysis gas, pyrolysis liquid, steam from incineration, and other forms of energy[9] are much higher than those discussed here, but insufficient detail is provided to evaluate this information. A more detailed set of estimated costs for dry-shredded-fuel processing plants has been prepared by the U.S. Environmental Protection Agency,[10] showing the effect of estimated revenue ranges, capacity utilization, and special costs such as taxes, transportation, high residue disposal charges, unusual site work, etc. These cost data show that such a project can range from profitable to a high cost of solid waste disposal, depending on the specific project conditions encountered.

Table 47. Range of Potential Values for Resource Recovery in Thermal Processing Facilities

Resource	Recoverable Units per Metric Ton of Mixed Waste	Realized Value FOB Plant, $/Unit			Value, $ per Metric Ton of Mixed Waste		
		A	B	C	A	B	C
1. Glass	0.05 MT[a]	5	20	50	0.25	1.00	2.50
2. Ferrous Metal (feed)	0.06 MT	5	20	50	0.30	1.20	3.00
3. Ferrous Metal (residue) (residue)[b]	0.04 MT	5	20	50	0.20	0.80	2.00
4. Nonferrous Metal	0.004 MT	150	250	400	0.60	1.00	1.60
5. Steam	3.3 MT	1.00	2.00	4.00	3.30	6.60	13.20
6. Electrical Power	450 KWH	0.008	0.02	0.03	3.60	9.00	13.50
	600 KWH	0.008	0.02	0.03	4.80	12.00	18.00
7. Fuel[c]	0.24 MT	10	25	40	2.40	6.00	9.60
($/$10^6$ BTU)		(0.43)	(1.08)	(1.73)			

[a]MT = metric ton = 2205 lb.

[b]Incinerator residue.

[c]Pyrolysis liquid at 5.834 x 10^6 Kcal/MT fuel (10,500 BTU/lb).

Table 48. Front End Separation Processes–Incremental Capital and Operating Costs[6]
Basis: 331,000 MT/yr Plant, 365 day/yr operation

Unit Operation[a]	Capital Cost[b] $/MT/hr ($/ST/day)	Ownership Cost[c] $/MT	Operating Cost $/MT	Total $/MT
Primary Shredding	33,070 (1250)	0.41	2.78	3.19
Air Classification	25,100 (950)	0.31	1.50	1.81
Secondary Shredding	16,500 (625)	0.20	1.76	1.96
Magnetic Metal Separation	2,000 (75)	0.02	0.43	0.45
Rising Current/Heavy Media Separation	6,900 (260)	0.08	0.45	0.53
Roll Crushing and Electrostatic Separation	7,400 (280)	0.09	0.49	0.58
Color Sorting	11,200 (425)	0.14	0.44	0.58

[a]Those listed are not all fully proven in large-scale operation.
[b]Does not include land or buildings.
[c]15-year amortization, 7%/yr interest.

Table 49. Economic Potential for Thermal Processing Facilities with Resource and Energy Recovery

	Ownership and Operating Costs $/MT[a]	Resource Recovery[b]	Resource Credits $/MT	Net Cost (Profit) $/MT
Incineration Only	8.47	None	–	8.47
Incineration and Residue Recovery	9.88	Ferrous and nonferrous metals, glass (1,3,4)	1.05-6.10	3.78-8.83
Incineration and Steam Generation	11.44	Steam (5)	3.30-13.20	(1.76)-8.14
Incineration with Steam Generation and Residue Recovery	12.89	Ferrous and nonferrous metal, glass steam (1,3,4,5)	4.35-19.30	(6.41)-8.54
Pyrolysis with Resource and Oil Recovery	12.08	Ferrous and nonferrous metals, glass, oil (1,2,4,7)	3.55-16.70	(4.62)-8.53

[a]For 272,000 metric ton (MT) per year facility; derived from a 1973 report to the President's Council on Environmental Quality.[8] To be used for orientation only.
[b]Numbers under "Resource Recovery" heading refer to items on Table 47.

Projected economics for two recent thermal processing facilities, summarized in Table 50, show that economically attractive projects are possible. Since both projects are still in their start-up phase, these results have yet to be verified.

Table 50. Projected Economics for Recent Energy Recovery Projects

Type of Project and Capacity	Incinerator With Steam Generation[a] 4 x 227 MT/day	Pyrolysis 907 MT/day[b]	
Steam Price, $/MT steam	$2.09	$1.79[b]	$4.83[b]
Operating Costs, $/MT	$3.40	$6.46	$6.46
Ownership Costs, $/MT	6.00	4.10	4.10
Total Costs, $/MT	$9.40	$10.56	$10.56
Credits from Sale of Steam, $/MT	6.28	4.29	11.59
Net Cost (Profit), $/MT of Waste Processed	$3.12	$6.27	($1.03)

[a]Original projected economics for 1980. Steam price to vary by formula as Bunker C price varies.

[b]Projected economics for 1975. $1.79/MT steam (81¢/1000 lb) based on $3.70/ barrel No. 6 fuel oil. Escalation in sales contract of $0.002189/1000 lb steam per 1¢ change in fuel oil price brings price to $4.83/MT ($2.19/1000 lb) for $10/barrel No. fuel oil.

REFERENCES

1. Zausner, E. R. An Accounting System for Incineration Operations. Public Health Service Publication No. 2022. Bureau of Solid Waste Management Report SW-17ts. U.S. Department of Health, Education, and Welfare. 1970. 17 pages.
2. Neissen, W. R. *et al.* Systems Study of Air Pollution From Municipal Incineration. Volume I. Arthur D. Little, Incorporated. Cambridge, Massachusetts. U.S. Department of Health, Education, and Welfare. National Air Pollution Control Administration Contract No. CPA-22-69-23. NTIS Report PB 192 378. Springfield, Va. March 1970. Pages VII-89-172.
3. Ricci, L. J. CE Cost Indexes Accelerate 10-Year Climb. Chemical Engineering. April 28, 1975. Pages 117-118.

4. Achinger, W. C. and Daniels, L. E. An Evaluation of Seven Incinerators. U.S. Environmental Protection Agency. Publication SW-51ts. lj. May 12-20, 1970. 76 pages.
5. Aubin, H. The New Quebec Metro Incinerator. Proceedings, 1974 National Incinerator Conference. Miami. May 12-15, 1974. American Society of Mechanical Engineers. Pages 203-212.
6. Schulz, H. W. Cost/Benefits of Solid Waste Reuse. Environmental Science & Technology **9**(5): 423-427. May 1975.
7. Resource Recovery—Catalogue of Processes. Prepared for the U.S. Council on Environmental Quality. Midwest Research Institute. National Technical Information Service. Springfield, Va. PB 214 148. February 1973. 141 pages.
8. Resource Recovery—The State of Technology. Prepared for the U.S. Council on Environmental Quality. Midwest Research Institute. National Technical Information Service. Springfield, Va. PB 214 149. February 1973. 67 pages.
9. Fuels from Municipal Refuse for Utilities. Technology Assessment. Prepared for Electric Power Research Institute. Bechtel Corporation. National Technical Information Service. Springfield, Va. PB 242 413. March 1975. 184 pages.
10. Third Report to Congress. Resource Recovery and Waste Reduction. SW-161. Office of Solid Waste Management Programs. U.S. Environmental Protection Agency. Washington, D. C. 1975. 96 pages.

CHAPTER 6

AIR POLLUTION CONTROL

Historically, the air pollution problems caused by incineration have been so severe that, in the public eye, "smoke" represents the single most associated image of incineration. However, with current-day technology and stringent federal, state and local regulations, modern, well-designed incinerators and other thermal processing facilities can be and are socially and environmentally acceptable.

Planning, design, specification, purchase, installation, operation and maintenance of air pollution control systems require at least as much attention as the furnaces, buildings, and other sections of the facility. Even at this late date in the evolution of regulations and technology for proper control, several recent projects have had major difficulties with air pollution control systems, some requiring major modifications or complete replacement. Particulate emissions are generally of greatest concern, but chemical emissions will also be considered in this discussion.

Most of the information presented here pertains to incinerators, both with and without energy recovery. Special sections are devoted to emerging thermal processing systems such as combined refuse/fossil fuel boilers and pyrolysis.

UNCONTROLLED PARTICULATE MATTER EMISSIONS

Any general discussion of particulate emissions may create confusion, since there is no universally accepted definition of or measurement procedure for "particulate." In an actual situation, careful study of applicable regulations and stack testing are essential. Examples of definitions and test procedures which exist for regulating particulate emissions from municipal incinerators are shown in Table 51.

Table 51. Typical Particulate Control Regulations for Incineration

Regulating Body	Definition of "Particulate"	Test Procedure	Remarks
U.S. Environmental Protection Agency[1]	Any finely-divided liquid or solid material, other than uncombined water, as measured by Method 5	U.S. EPA Method 5	Regulates all matter which is collected on a filter at ≥121.1°C (250°F)
N.J. Department of Environmental Protection[2]	Any material, except uncombined water, which exists in a finely divided form as liquid particles or solid particles at standard conditions [21.1°C (70°F), 1 atmosphere absolute pressure].	None specified	Regulates all matter which is particulate at ≥21.1°C (70°F)
County of Los Angeles Air Pollution Control District	Any material, except uncombined water, which exists in a finely divided form as a liquid or solid at standard conditions [15.6°C (60°F), one atmosphere absolute pressure].	None specified	Regulates all matter which is particulate at ≥15.6°C (60°F)
Virginia State Air Pollution Control Board	Any material, except water in uncombined form, that is airborne and exists as a liquid or a solid at standard conditions [21.1°C (70°F), one atmosphere absolute pressure].	ASME PTC-27[a] or IIAT-6[b]	Regulates all matter which is particulate at ≥21.1°C (70°F), but tests for that which is particulate at stack temperature.

[a]American Society of Mechanical Engineers Power Test Code 27.
[b]Incinerator Institute of America Bulletin T-6, "Incinerator Testing."

For purposes of simplicity, particulate emissions which are filterable at approximately 121.1°C (250°F) will be referred to as "dry catch"; particulates which pass through the filter will be referred to as "wet catch"; **total** particulates equal dry catch plus wet catch. Unless otherwise noted, the data presented in this chapter will be presumed to be "dry catch" only.

Particulate Emission Quantities

A 1970 study[3] compiled particulate emission data from various locations downstream of the furnaces at 50 different incinerators which had no air pollution control devices. These data are shown in Figure 25, a histogram showing the wide variation that exists between incinerators, and even for the same incinerator. A median value of 12 kilograms of particulate emissions per metric ton of refuse (24 lb/short ton) was determined. This compares reasonably well with median values of 8.5, 17.5, 10, and 11.3 kilograms per metric ton reported by others.[4-8] A detailed study of a modern waterwall steam-generating incinerator showed similar emissions upstream of the pollution control device (Table 52).

Often, regulations are written to limit particulate emission concentrations to a specified value based on operating the incinerator with an amount of air which will result in a carbon dioxide content of 12% by volume in the flue gas (excluding the contribution of auxiliary fuel), when measured on a moisture-free basis. Therefore, the Table 52 data are presented in other forms in Table 53. Reference 3 contains factors for conversion to additional forms. It should be noted that many regulations define "normal" or "standard" differently from historical usages. Therefore careful reading of definitions is essential.

Uncontrolled particulate emissions vary, depending on the construction and operation of the equipment, as well as the nature of the waste. Some of the major variables are:

ash content of the solid waste
underfire and overfire air flows
burning rate
furnace temperature
grate agitation
combustion chamber design

Three mechanisms are believed to be mainly reponsible for particulate emissions:

mechanical entrainment of particles from the burning waste bed
the cracking of pyrolysis gases
the vaporization of metal salts or oxides

An extensive and detailed discussion of these variables and mechanisms are presented in Reference 3. It should be noted that while the particulate emissions from a waterwall furnace may be similar to that from a conventional furnace on a per ton of refuse basis, waterwall emissions may be significantly higher on a concentration basis, due to lower air flows (See Chapter 2).

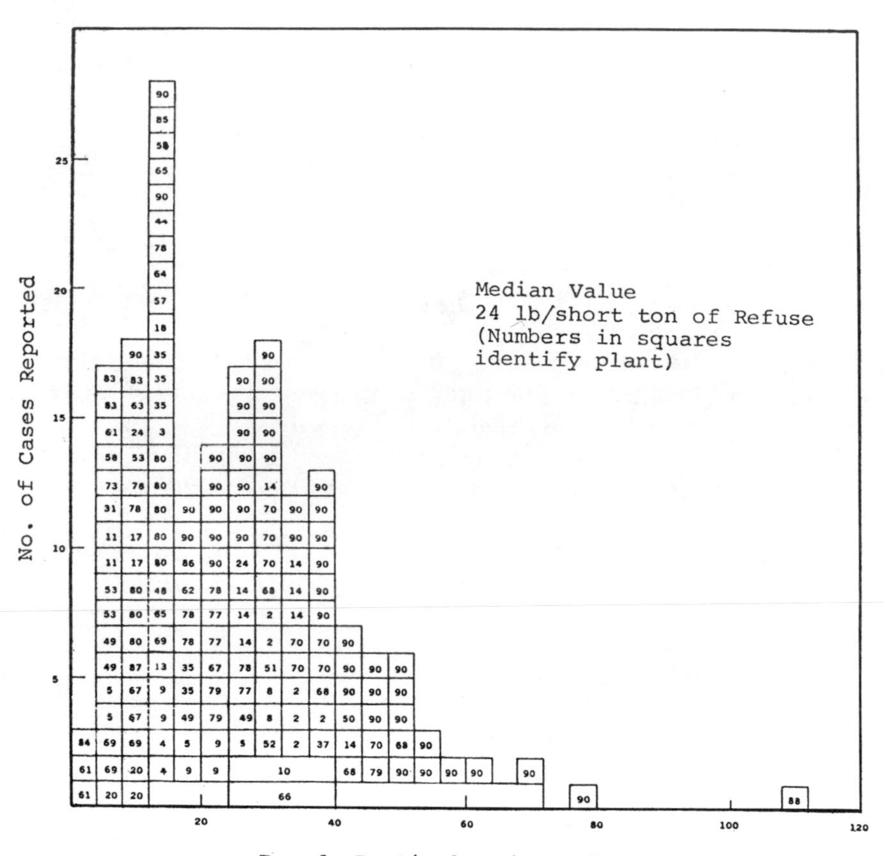

Figure 25. Histogram of particulate furnace emission factors for municipal incinerators (capacity: >50 tons/day).

Table 52. Particulate Emissions from the Furnace of a Modern Waterwall Incinerator[9]

Refuse Charging Rate, short tons/hr	16.6	16.6	16.7	16.7
Volume % CO_2 in Flue Gas (Drv Basis)	10.0	10.0	10.1	9.5
Dry Catch Particulates, lb/hr	388	379	427	398
Wet Catch Particulates, lb/hr	30	18	30	13
Total Particulates, lb/hr	418	397	457	411
Dry Catch Particulates, lb/short ton	23.4	22.8	25.6	23.8
Wet Catch Particulates, lb/short ton	1.8	1.1	1.8	0.8
Total Particulates, lb/short ton	25.2	23.9	27.4	24.6
Total Particulates, kg/metric ton	12.6	12.0	13.7	12.3

Table 53. Particulate Emissions from the Furnace of a Modern Waterwall Incinerator[9]

Refuse Charging Rate, short tons/hr	16.6	16.6	16.7	16.7
Excess Air, %	78	78	87	98
Volume % CO_2, dry basis	10.0	10.0	10.1	9.5
Volume % H_2O	11.0	10.8	13.9	12.7
Total Particulates				
Grains/SCF (dry), actual	1.05	1.10	1.03	0.93
Grains/SCF (dry), corr. to 12% CO_2	1.26	1.32	1.22	1.17
Grams/SCM (dry), corr. to 12% CO_2	2.88	3.02	2.79	2.68

Grains = 1/7000 Pound = 1/15.43 grams

SCF (dry) = standard cubic feet of dry flue gas, @ 21.1°C (70°F) one atmosphere absolute pressure
= 0.0283 standard cubic meters (SCM) dry @ 21.1°C (70°F) one atmosphere absolute pressure

Characteristics of Particulates

The composition of particulate emissions from incinerator furnaces is dependent on design and operation, as well as on the refuse ash composition. A poorly designed or operated incinerator may emit carbon particles (usually referred to as soot), and the inorganic (mineral) type ash will contain a significant quantity of combustibles. Data from six incinerators[3] showed a range of 6-40% in the combustible content of the furnace particulate emissions. Inorganic contents are shown in Table 54.

Table 54. Composition of Inorganic Components of Particulates from Furnaces[3]

Component	Computed for Typical Refuse	NYC Incinerators[10]	
		73rd St.	So. Shore
SiO_2	53.0%	46.4%	55.1%
Al_2O_3	6.2	28.2	20.5
Fe_2O_3	2.6	7.1	6.0
CaO	14.8	10.6	7.8
MgO	9.3	2.9	1.9
Na_2O	4.3	3.0	7.0
K_2O	3.5	2.3	–
TiO_2	4.2	3.1	–
SO_3	0.1	2.7	2.3
P_2O_5	1.5	–	–
ZnO	0.4	–	–
BaO	0.1	–	–
	100.0%		

Particle size distribution and specific gravity of particulate matter are properties which are essential design data for most particulate removal devices. The smaller and/or finer particles require more sophisticated (and expensive) equipment to meet a specific emission limit. Table 55 presents data for three conventional incinerator furnaces. Due to less efficient particle collection testing methods used in the past, these data could be in error in that a portion of the fine particles were not included.

Figure 26 presents additional particle size distribution data. Size distribution, like loading, varies widely. Most factors which affect

Table 55. Properties of Particulates Leaving Furnaces[11]

Physical Analysis	Installation 1 (250 TPD)	2 (250 TPD)	3 (120 TPD)
Specific Gravity, g/cc	2.65	2.70	3.77
Bulk Density g/cc (lb/CF)	–	0.495 (30.9)	0.151 (9.4)
Loss on Ignition at 750°C, wt %	18.5	8.15	30.4
Size Distribution (% by weight)			
2 microns	13.5	14.6	23.5
4 microns	16.0	19.2	30.0
6 microns	19.0	22.3	33.7
8 microns	21.0	24.8	36.3
10 microns	23.0	26.8	38.1
15 microns	25.0	31.1	42.1
20 microns	27.5	34.6	45.0
30 microns	30.0	40.4	50.0
Particulate Emission Rate, kilograms/MT (lb/ST)	6.1(12.1)	12.3(24.6)	4.6(9.1)

TPD = short tons per day
MT = metric ton
ST = short ton

particle loading also affect the size of particles emitted. Improved incinerator performance which reduces quantities emitted normally decreases the size of the individual particles. The particulate matter is always quite heterogeneous, consisting of flyash, with properties such as shown in Tables 54 and 55, combined with large, low-density flakes. Particle density typically ranges from 2-3 g/cc.

Electrical resistivity is an important property of particulates necessary for design of electrostatic precipitators, commonly used in modern incinerators. High resistivity reduces collection efficiency, while low resistivity may result in re-entrainment of the particle into the gas stream after collection. Resistivity is a function of the basic particle characteristics, and composition and temperature of the flue gas stream. The presence of moisture and very low concentrations of certain chemical compounds in the flue, such as sulfur trioxide and ammonia gas, may strongly influence particle resistivity and precipitator efficiency. Figure 27 shows the particle electrical resistivity for emissions from three furnaces. The desirable range of resistivity, 10^4-10^{10} ohm-cm, influences the choice of electrostatic precipitator operating temperature.

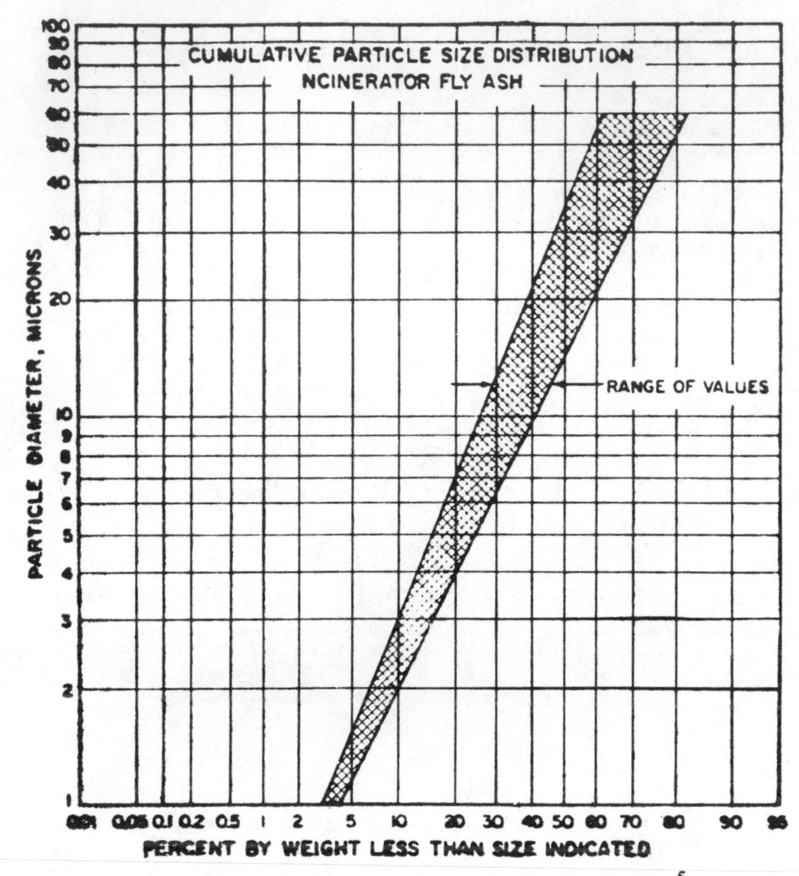

Figure 26. Incinerator flyash particle size distribution.[5]

TARGET PARTICULATE EMISSION LEVELS

Allowable particulate emissions are determined by three standards, usually concurrently:

- Air quality in the regions affected by the thermal processing facility.
- Concentration or rate of emissions from the thermal processing facility.
- Visual appearance of the emissions from the thermal processing facility.

These standards exist on the federal level for new facilities (construction commenced after 12/23/71),[1] and on the state level for existing facilities as well.

Federal (and many state) requirements for air quality are set at two levels, as shown in Table 56, a primary standard which is designed to protect public health, and a secondary standard which is designed to protect public welfare (*e.g.*, animal or plant life, or property, or

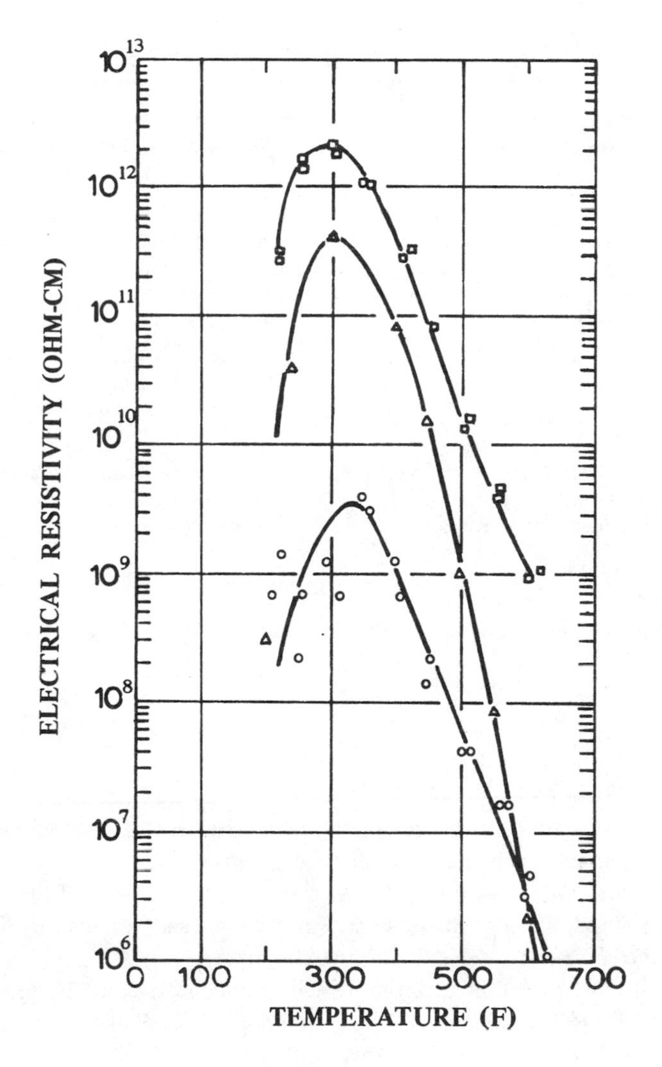

Figure 27. Bulk electrical resistivity of entrained particulates leaving three large, continuous-feed furnaces at 6% water vapor.[11]

enjoyment thereof). For purposes of air quality standards, "particulate" generally means that which is filterable from the air and which remains on the filter after conditioning at 15 35°C (59-95°F).

With the sophisticated computer modeling techniques now available, the air quality in regions affected by a thermal processing facility can be reasonably predicted for varying stack emission rates and meteorological conditions (*e.g.*, wind speed and direction, vertical temperature profiles,

Table 56. National Ambient Air Quality Standards[12]

National Primary Ambient Air Quality Standards for Particulate Matter	National Secondary Ambient Air Quality Standards for Particulate Matter
The national primary ambient air quality standards for particulate matter, measured by the reference method described in the regulation[a] or by an equivalent method, are:	The national secondary ambient air quality standards for particulate matter, measured by the reference method described in the regulation[a] or by an equivalent method are:
(a) 75 $\mu g/m^3$–annual geometric mean	(a) 60 $\mu g/m^3$–annual geometric mean, as a guide to be used in assessing implementation plans to achieve the 24-hr standard.
(b) 260 $\mu g/m^3$–maximum 24-hr concentration not to be exceeded more than once per year	(b) 150 $\mu g/m^3$–maximum 24-hr concentration not to be exceeded more than once per year

[a]High volume sampling method described in Appendix B of reference 12.

humidity, etc.), taking into account stack height and exit velocities, land topography, already-existing particulate concentrations, and other sources of particulates. While not specific to thermal processing facilities, the air quality standards nevertheless must not be exceeded because of insufficient control in new and existing installations.

For thermal processing facilities which commenced construction after December 23, 1971, and which charge more than 1.89 metric tons per hour (50 short tons/day) of solid waste,* the very specific federal regulations for particulate emission concentrations must be met.[1] These standards prohibit the discharge into the atmosphere of particulate matter, the concentration of which is in excess of 0.18 grams per cubic meter (@ 21.1°C, one atmosphere) which is equal to 0.08 grains/standard cubic foot (21.1°C, one atmosphere) of flue gas on a dry basis corrected to 12% carbon dioxide by volume, maximum 2-hr average. The particulate emissions are to be measured in accordance with U.S. Environmental Protection Agency "Method 5, Determination of Particulate Emissions

*Defined as refuse, more than 50% of which is municipal type waste.[1]

From Stationary Sources,"[1] which measures only "dry catch" particulates. A requirement for recording burning rates, hours of operation, and any particulate emission measurements which are made is also included.

It was originally proposed that the federal emission standards for incinerators and other sources would be for total particulates, which included dry catch and wet catch. As can be deduced from Tables 52 and 53, for incinerators the wet catch corresponds to a quantity on the same order of magnitude as the emission standard. As will be seen in the control section, the wet catch particulates are not efficiently collected by some devices and in themselves could cause noncompliance if this proposed standard were ever promulgated.

On the state level, similar regulations exist for both new and existing facilities, but the actual numerical standard, the definition of "particulate," and the test method may be significantly different. State regulations may also have prohibitions against particulate emissions which are visible. These regulations may set a specific standard such as a limit on opacity* (typically 20%), or an equivalent smoke density (*e.g.,* Ringelmann No. 1), or simple prohibition of any visible particles.[2]

Actual data correlating particulate concentration with the resulting opacity or smoke density have not been found, but a correlation is presented in Figure 28 using a method[13] which takes into account particulate concentration, refractive index, geometric mean radius (and standard deviation), density, stack diameter, and gas temperature. The claim that 0.09 grams per standard cubic meter (0.04 grains/standard cubic foot) results in a clear stack[14] is not inconsistent with this correlation.

PARTICULATE EMISSIONS CONTROL

Even a modern, well-designed and operated incinerator cannot meet federal and most, if not all, state regulations for particulate emissions without an air pollution control system. Comparing data from Table 53 with federal emissions requirements (Table 57), it is apparent that efficiencies in excess of 93% on a weight basis are required. Visual requirements, by state or local agencies, of less than 20% opacity may increase this efficiency requirement even further.

Mass particulate emission standards are corrected to 12 volume % CO_2 (excluding the contribution of auxiliary fuel) or some other measure of excess air, which effectively limits the emission to a fixed amount per ton of solid waste fired. Thus, air dilution in a refractory incinerator, which may use twice as much air as a waterwall incinerator, is not an

---

*The degree of obstruction to the transmission of light.

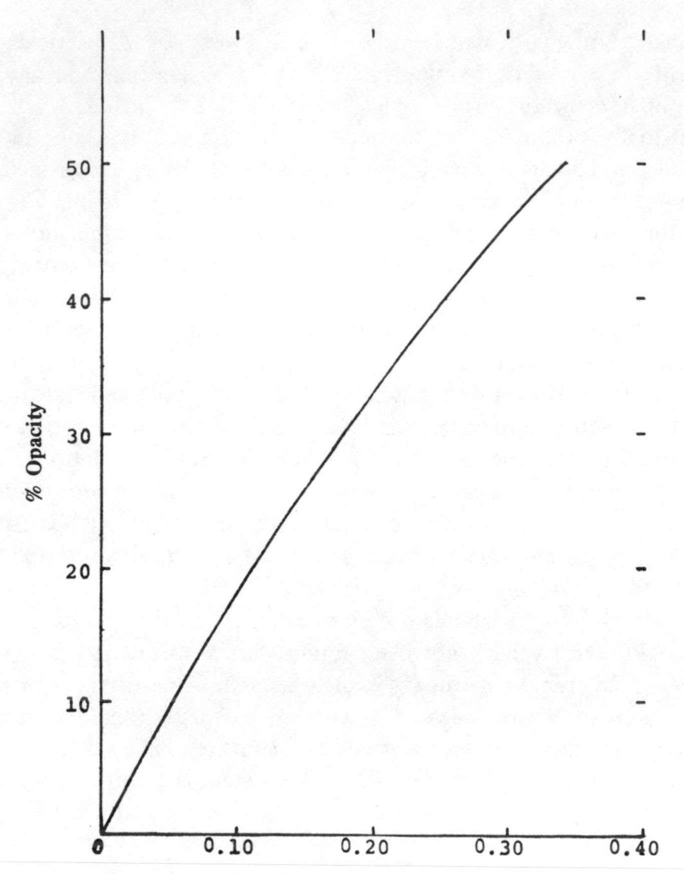

Actual*Particulate Concentration, grains/standard cubic foot (wet basis)

*Not corrected to 12% CO_2/ + 21.1°C, 1 atmosphere

ASSUMPTIONS:

1. Geometric mean particle radius = 0.85 μ.
2. Geometric standard deviation for particle distribution = 1.59.
3. Particulate is silica, with density of 2.5 g/cc, refractive index of 1.50.
4. Stack diameter = 1.83 m (6 ft).
5. Stack gas temperature = 260°C (500°F).
6. (Grams per standard cubic meter) = (grains per standard cubic foot) x 2.29.

Figure 28. Estimated correlation between opacity and particulate concentration.

Table 57. Particulate Emission Data from Uncontrolled Waterwall Incinerator Compared with Federal Standards

Particulate Emission Concentration	Uncontrolled Incinerator[a]	Federal Standard or Equivalent	% Reduction Required
Grains/SCF (dry)[b]	1.18	0.08	93.3
Grams/SCM (dry)[c]	2.70	0.18	

[a]Derived from Tables 52 and 53; average of four tests corrected to 12% CO_2 by volume, dry basis; "dry catch" only.

[b]Standard (70°F, 29.92 in. Hg) cubic feet, dry basis.

[c]Standard (70°F, 29.92 in. Hg) cubic meters, dry basis.

70°F = 21.1°C   29.92 in. Hg = 1 atmosphere

aid in meeting emission limits. However, because opacity is an absolute standard, the refractory incinerator may in fact be aided by normal dilution in meeting an opacity requirement. It may also be aided by decreasing the size of a single stack, *e.g.*, by using four stacks with four incinerator trains instead of one or two stacks, because of the effect of stack size on opacity measurements.

Considering the particle size data presented in Figure 26, it is apparent that to achieve a minimum of 90% efficiency, all the particles larger than 1-3 μ (one-millionth of a meter) must be removed. This requirement effectively eliminates the simple air pollution control systems traditionally used on incinerators, although it may sometimes be advantageous to use one of these simpler devices as a first-stage collector, for example, to reduce the required efficiency of the final collector. Numerous discussions of these systems, which include settling chambers, wetted baffle spray systems, cyclones, and low-energy scrubbers are available.[5,15,16] Therefore, they will not be considered further in this publication.

Electrostatic precipitators, fabric filters, and certain types of scrubbers appear to be the only commercially-available devices which have the capability to meet the current emission standards for municipal incinerators. Newer forms of these devices, including charged droplet scrubbers and high-velocity wet precipitators, may have advantages over more conventional devices but these have not been commercially demonstrated for incinerator applications.

Electrostatic Precipitators

Electrostatic precipitators have been used in utility and industrial steam-generating boilers and many other applications for over 50 years with a relatively good performance record. Not until 1969 were these devices applied to municipal incinerators in the U.S., although there are probably more than 40 installations in Europe and Japan. Almost all new thermal processing facilities built since 1969, however, have utilized electrostatic precipitators for particulate emission control (Table 58).

In an electrostatic precipitator, a high, normally negative voltage gradient is impressed across a pair of electrodes producing a corona discharge (the visible sign of ionization of gas molecules) at the negative electrode. Most of the ions produced are negatively charged. As the ions migrate toward the grounded (relatively positive) electrode, they collide with entrained particles, charging these particles negatively. The negatively charged particles in turn move toward the grounded electrode where they are attached and held by a combination of electrical, adhesive, and cohesive forces while their negative charge is gradually conducted through the layers of previously collected dust to the grounded electrode. The resistance to conduction is termed "dust resistivity."

Too high a resistivity results in a high voltage drop across the dust layer, reducing particle collection because of depressed electrical fields in the precipitation zones and lower levels of particle charge.[18] In extreme cases, very high resistivity may result in "back corona," generating positive ions which tend to neutralize the electrical field and upset particle collection. Resistivity, which is a complex function of both gas and particle characteristics, is a primary parameter to be considered in precipitator design.

The major elements of a commercial electrostatic precipitator are the high voltage power supply and controls, discharge (corona) electrodes, collecting (grounded) electrodes, rappers to dislodge agglomerated dust from electrode surfaces, a gas-tight shell to contain the precipitation zone, hoppers below the precipitation zone to receive dislodged dust, and gas inlet and discharge zones designed to distribute the gas uniformly across the precipitation zone. A typical configuration is shown in Figure 29.

High-Voltage Power Supply and Controls

The high-voltage system is designed to provide high-voltage direct current to the discharge electrodes. The system consists of:

1. Typically 460-volt, 60-Hz, single-phase alternating current power supply.
2. An electrical control circuit incorporating either a saturable reactor or an SCR (silicon-controlled rectifier).

Table 58. Partial Listing of Electrostatic Precipitator Installations at Thermal Processing Facilities in the U.S. and Canada, Including Design Parameters[a]

Plant	Capacity TPD	Furnace Type[b]	Gas Flow ACFM	Gas Temp °F	Gas Velocity FPS	Residence Time Sec	Plate Area ACFM/ft^2	Input KVA	Pressure Drop in. H_2O gage	Efficiency wt %
Montreal	4 x 300	WW	112,000	536	5.5	3.3	6.2	35	0.5	95.0
Stamford	1 x 220	Special R	160,000	600	6.0	3.3	6.6	57	0.5	95.0
Stamford	1 x 360	R	225,000	600	3.6	5.0	4.5	225	0.5	95.0
Stamford	1 x 150	R	75,000	600	3.7	4.9	4.6	75	0.5	95.0
SW Brooklyn	1 x 250	R	131,000	550	4.4	3.2	6.7	47	2.5	94.3
So. Shore, N.Y.	1 x 250	R	136,000	600	5.5	3.3	6.8	33	0.5	95.0
Dade City, Fla.	1 x 300	R	286,000	570	3.9	4.0	5.7	48	0.4	95.6
Chicago, NW	4 x 400	WW	110,000	450	2.9	4.6	5.5	40	0.2	96.9
Braintree, Mass.	2 x 120	WW	32,000	600	3.1	4.5	5.5	19	0.4	93.0
Hamilton, Ont.	2 x 300	WW	81,000	585	3.5	5.4	3.9	70	0.5	98.5
Washington, D.C.	6 x 250	R	130,800	550	4.1	3.9	4.9	77	0.4	95.0
Eastman Kodak	1 x 300	WW	101,500	625	3.4	5.5	3.8	106	—	97.5
Harrisburg, Pa.	2 x 360	WW	100,000	410	3.5	5.1	5.0	40	0.2	96.8

[a]Most of data from reference 17.

[b]R = refractory-lined; WW = waterwall.

NOTE: Except for capacity, data refer to design parameters for one precipitator; several may exist.

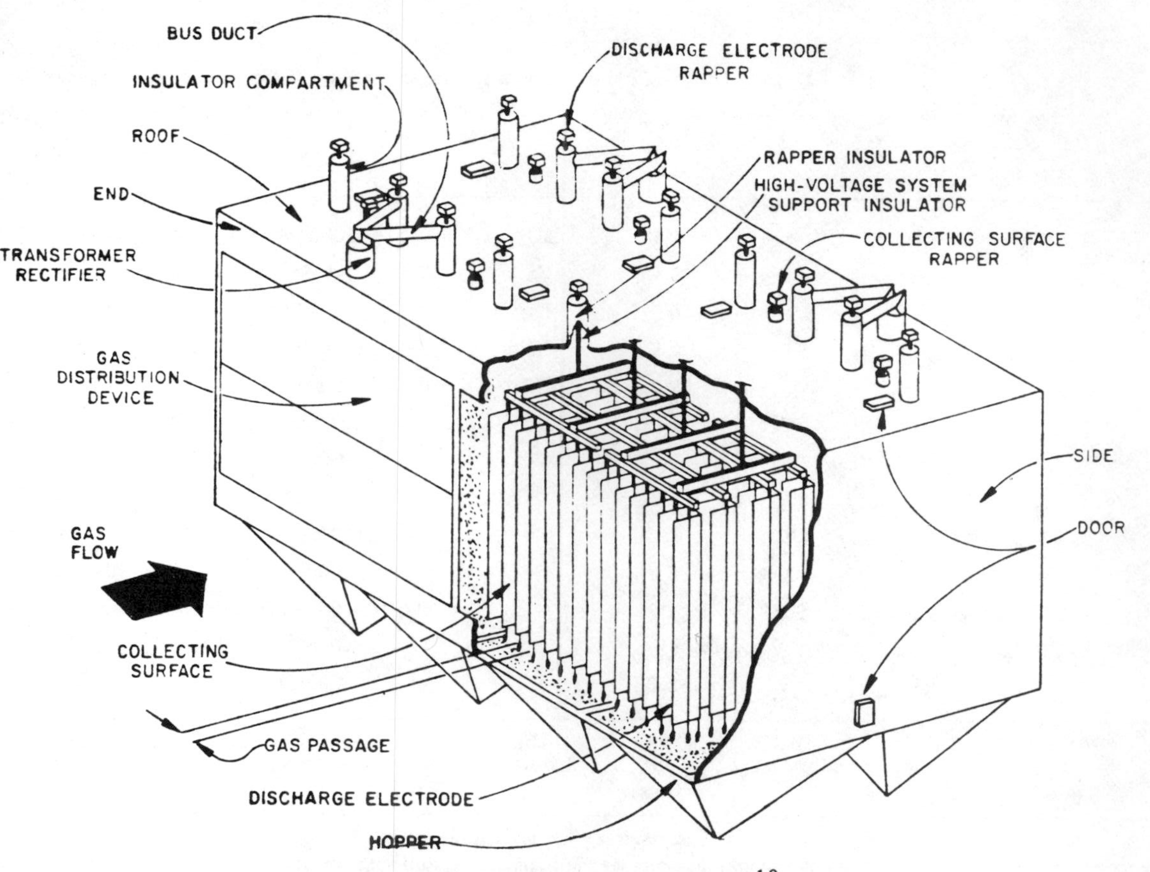

Figure 29. Typical electrostatic precipitator.[19]

3. A transformer-rectifier set (encased in an oil tank) to increase the voltage to the desired value in the range of 30,000 to 80,000 **volts**, and to convert alternating to direct current. Silicon diode rectifiers have replaced earlier mechanical, vacuum tube, and selenium types.

Automatic control systems and electrical sectioning of precipitators, although increasing initial cost somewhat, are important in maintaining maximum precipitator performance.[20] Proper design and placement of high-voltage insulators is necessary to overcome failures due to dirt and moisture.

Discharge Electrodes

The discharge electrodes, which take the form of rods or wires with or without superimposed sharp points or edges to enhance the corona effect, hang precisely along the center lines of the gas passages between the collection electrodes. The discharge electrodes must be designed with great care to maintain uniform spacing and to minimize fatigue. These may be the suspended wire type with weights, as shown in Figure 29, or the electrodes may be mounted in stiff frames. Each electrical section, containing a multitude of discharge electrodes, is electrically isolated from the remainder of the precipitator.

Collection Electrodes

For incinerator applications, these are parallel steel plates carefully spaced at a predetermined value in the range of 20-30 cm (8-12 in.) apart, depending on the voltage used. The plates may be smooth, but are normally corrugated or equipped with fins or carefully designed baffles to increase strength and to provide quiescent zones, avoiding particle re-entrainment while maintaining smooth gas flow.

Rappers

Rappers or vibrators are used during precipitator operation to shear the collected dust layers away from the collecting electrode. Both intensity and time may be controlled to induce shearing off relatively large agglomerates, avoiding particle re-entrainment. Such losses are also minimized by sequential rapping of shock-isolated rapping sections in series, so that an opportunity exists to recapture particles which are re-entrained in the inlet sections. A variety of rapping mechanisms and locations are used, including mechanical, electromagnetic, and pneumatic devices, top- or end-mounted. A group of mechanically-driven, free-swinging hammers on a single shaft, each hammer rapping the end of single plate, is one simple arrangement, but one in which intensity is not readily controlled. Similar rapping arrangements are adapted for use on groups of discharge (negative) electrodes because some dust almost inevitably clings to these, due to impaction and positive ion formation.

Flyash Removal

The hoppers may be discharged periodically through slide valves or continuously through rotary valves or feeders to a flyash removal system. The removal system is usually dry, using screw or drag conveyors, or vacuum or pneumatic transport. However, slurry systems can be used.

Proper hopper design and dumping procedures are vital, since buildup of flyash can short out electrodes causing precipitator damage. Level detectors are useful, and automatic timing devices should be used where periodic dumping is practiced.

Gas Inlet and Discharge Zones

Uniform flow into and through the precipitator must be maintained to insure adequate performance. The reduction in gas velocity from the usual practice of about 10-20 m/sec (33-66 ft/sec) in inlet ducts to about 0.9-1.8 m/sec (3-6 ft/sec) in the precipitator, changes in gas direction, and the necessity for limiting the amount of total ductwork pose difficult design problems with regard to obtaining uniform flow. Careful design of splitters or vanes and incorporation of perforated plates at the precipitator inlet will improve gas distribution. Test models scaled 1 to 16 are sometimes used to aid in design work. Flow distribution tests should be included in performance evaluation for new or rebuilt precipitators.

Gas distribution through the precipitator is affected too by the gas outlet design, also deserving of careful attention. Another serious problem can be created by gas by-passing below the collecting plates through the hopper area. This condition can usually be improved by use of baffles.

Precipitator Sizing

The design of electrostatic precipitators is based on knowledge of gas and solid physical and chemical properties, particle inlet loadings, gas rate, required efficiency, and also on familiarity with the idiosyncracies of the particular application. For example, for incinerator operation the design should reflect knowledge of the expected temperature range, the frequency of shutdown, variability in flyash composition and properties, geographical location, and other variables. Using both theoretical and empirical design considerations, and considering cost optimization, the designer will specify major parameters, usually in the ranges shown in Table 59.

Table 59. Typical Electrostatic Precipitator Design Parameters for Incinerator Applications

Plate Spacing	20-30 cm (8-12 in.)
Velocity through Precipitator	0.9-1.8 m/sec (3-6 ft/sec)
Vertical Height of Plates	3.6-10 m (12-48 ft)
Horizontal Length of Plates[a]	0.5-1.5 x height
Applied Voltage	30,000-80,000 volts
Gas Temperature	177-343°C (350-650°F)
Gas Residence Time in Precipitator	3-6 sec
Draft Loss	3-20 mm water (0.1-0.8 in.)
Fields (electrical stages) in Direction of Gas Flow	1-4
Total Power for Precipitator	7-35 KW per m^3/min (0.2-1 KW/1000 ACFM)
Collection Area	400-1000 m^2 per 1000 m^3/min (122-305 ft^2/1000 ACFM)
Efficiency	93-99%
Gas Flow per Precipitator	850-8500 m^3/min (30,000-300,000 ACFM)
Migration Velocity[b]	6-12 cm/sec (0.2-0.4 ft/sec)

[a]Aspect ratio = [total horizontal length (depth) of collection plates] ÷ (height of collection plates) = 0.5-1.5.

[b]Mean average effective migration velocity of a particle toward the collection electrode. Sometimes called precipitation rate or drift velocity.

Although the required mass particle removal efficiency, based on federal emission standards previously discussed, may be less than 95%, opacity requirements approaching a "clear stack" may bring design efficiency to 99% (or even higher), depending on uncontrolled loading, particle size, condensibles present, stack diameter, and other factors. High design efficiencies require extraordinary attention to all design and construction details to insure continuing high efficiency performance. It should be emphasized that the mechanical and electrical designs, some important aspects of which have been discussed here, are as important to adequate electrostatic precipitator performance as the basic size parameters. Electrostatic precipitator installations both in the U.S. and elsewhere have shown that, with careful design and operation, efficiency requirements for "dry catch" particulates can be met. Actual performance data for an electrostatic precipitator operating on the effluent from a waterwall incinerator are provided in Table 60.

Table 60. Performance Data from Electrostatic Precipitator on Waterwall Furnace[9]

Refuse Charging Rate				
Short tons/hr	16.6	16.6	16.7	16.7
Dry Gas Composition (by volume)				
% CO_2	10.0	10.0	9.9	9.1
% O_2	9.4	9.4	7.4	8.5
Excess Air, %	78	78	87	98
Inlet Measurements				
Flow Rate, SCFM (dry)	46,500	42,100	51,900	51,500
Temperature, °F	338	307	400	415
Water Content, % by volume	11.0	10.8	13.9	12.7
Particulates–dry catch				
grains/SCF (dry) corrected[a]	1.17	1.26	1.14	1.14
grains/SCF (dry) actual	0.975	1.05	0.941	0.865
Particulates–wet catch				
grains/SCF (dry) corrected[a]	0.090	0.060	0.079	0.036
grains/SCF (dry) actual	0.075	0.050	0.065	0.027
Particulates–total				
grains/SCF (dry) corrected[a]	1.260	1.320	1.219	1.176
grains/SCF (dry) actual	1.050	1.100	1.006	0.892
Outlet Measurements				
Flow Rate SCFM (dry)	48,600	43,400	51,900	51,500
Temperature, °F	358	356	393	398
Water Content, % volume	9.7	8.5	13.9	12.7
Particulates–dry catch				
grains/SCF (dry) corrected[a]	0.0331	0.0283	0.0400	0.0270
grains/SCF (dry) actual	0.0276	0.0236	0.0330	0.0205
Particulates–wet catch				
grains/SCF (dry) corrected[a]	0.0103	0.0134	0.0090	0.0130
grains/SCF (dry) actual	0.0086	0.0112	0.0074	0.0099
Particulates–total				
grains/SCF (dry) corrected[a]	0.0434	0.0417	0.0490	0.0400
grains/SCF (dry) actual	0.0362	0.0348	0.0404	0.0304
Efficiency				
% Removal–dry catch	97.17	97.75	96.49	97.63
% Removal–wet catch	88.56	77.67	88.61	63.89
% Removal–total	96.56	96.84	95.98	96.60

[a]Corrected to 12% CO_2.

Corrosion due to acidic components of the flue gas, such as hydrogen chloride, can be a problem in precipitator operation. The gas temperature following heat and/or quench facilities must be sufficiently high to avoid acid condensation on cold surfaces. Hot air purging and preheat burners can minimize acid gas contact with cold surfaces during shutdown and startup. Sufficient insulation of all metal surfaces exposed to outdoor conditions is especially important. Hopper heaters are useful, both to avoid corrosion and to avoid bridging problems due to even minor amounts of moisture deposition on flyash.

Scrubbers

Devices that contact incinerator flue gas with water have traditionally been used both to clean the gases and to cool them for protection of duct, stack and fan materials. These devices, which usually consisted of little more than a large spray chamber with baffles, are inadequate to meet modern emission control requirements, though similar devices can still be used where flue gas cooling is required. As noted earlier, in order to meet most recent particulate emission regulations, it is necessary to install devices which efficiently remove particles in the 1-5 micron size range. This can be accomplished by some forms of a more sophisticated family of gas/water contacting devices known as scrubbers, or sometimes as wet or water scrubbers.

Various techniques are used in scrubbing, but all rely on "wetting" the particle with water in order to enlarge them, allowing for easier removal from the gas stream. The efficiency of a particular type of scrubber on a given particle size can be related to the energy used to force the gas through the collector and to generate the water sprays. This energy may be supplied with fans as pressure to the gas stream (gas-motivated), with pumps as pressure to the water stream (liquid-motivated) or mechanically. The latter has not found commercial usage, but the former methods are widely used.

Gas-motivated scrubbers are referred to as venturi or orifice types. Liquid-motivated scrubbers are referred to as jet venturi or ejector types, or impact scrubbers. The energy required in either case represents a very significant incinerator operating cost, especially when electrical power is used to drive fan or pump motors.

Research and development work is underway to improve the performance of scrubbers and to reduce energy requrements.[21] Various commercial claims are made that such energy reductions are already possible. While this may be so, it is beyond the scope of this publication to assess such developments which have not been demonstrated for thermal processing applications. However, any scrubber proven capable of performing equivalently to those described herein with lower energy consumption merits special consideration, because of the very significant contribution of scrubber energy losses to incinerator utility costs.

In the venturi-type scrubber, the gases are passed through a restricted "throat," where water is injected and the gas velocity accelerated, typically to 61-122 m/sec (200-400 ft/sec), promoting intimate gas-liquid contacting, with gas pressure drops up to 122 mm Hg (60 in. H_2O) and higher. The wetted particles are then collected in a mechanical-type device, such as a cyclonic separation chamber and/or wire screen demisters. Water rates are in the range of 0.7-5.3 CM/1000 CM (5-40 GPM/1000 CFM) of gas at the scrubber outlet (Table 61). Variations of the venturi or orifice gas-motivated scrubbers may be found in Figures 30, 31, and 32.

Table 61. Typical Venturi Scrubber Operating Parameters

Throat Gas Velocity		Water Rate[a]		Pressure Drop	
m/sec	ft/sec	CM/1000 CM	GPM/1000 CFM	mm Hg	in. H_2O
18-38	60-125	2.0-5.3	15-40	3.7-37	2-20
30-84	100-275	1.1-2.7	8-20	15-27	8-40
61-122+	200-400+	0.7-2.0	5-15	28-149+	15-80+

[a]Total water rate per unit rate of water-saturated gas. Usually, the major portion of the water requirement is supplied by recycled water.

Energy for a venturi scrubber is supplied by fans upstream or downstream of the scrubber. The power requirement is about 4 kilowatts per 1000 cubic meters/minute per mm Hg pressure drop (about 0.25 horsepower per 1000 cubic feet per minute per inch water pressure drop). Location of the fan is significant. A downstream fan must contend with increased mass flow due to water evaporation into the gas stream, although the volumetric flow rate may be lower because of the lower temperature. A downstream fan must also contend with impingement of sometimes corrosive water droplets, carried over from the scrubber or from condensation; while an upstream fan must contend with the design and maintenance problems associated with operation on a dirty gas at elevated temperatures. The downstream option is normally chosen for incinerator-type systems.

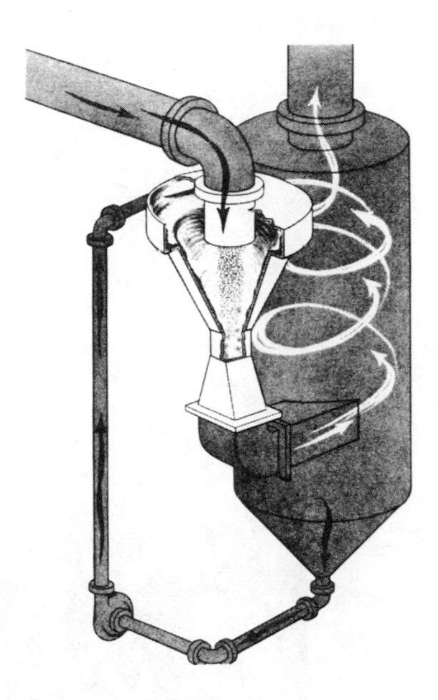

Figure 30. Gas-Motivated Venturi Scrubber Variation (Courtesy of Chemico).

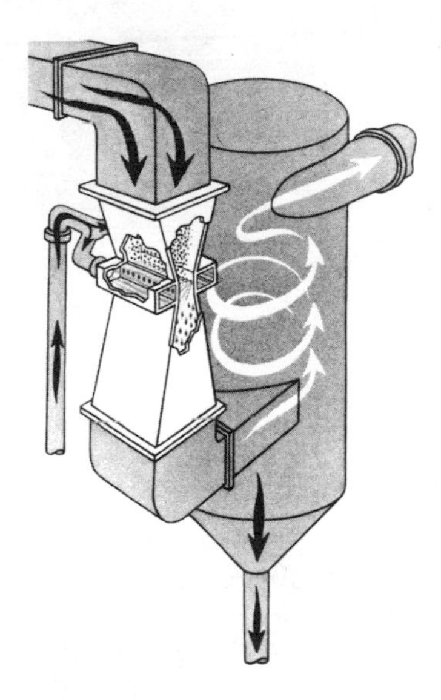

Figure 31. Gas-Motivated Venturi Scrubber Variation (Courtesy of Chemico).

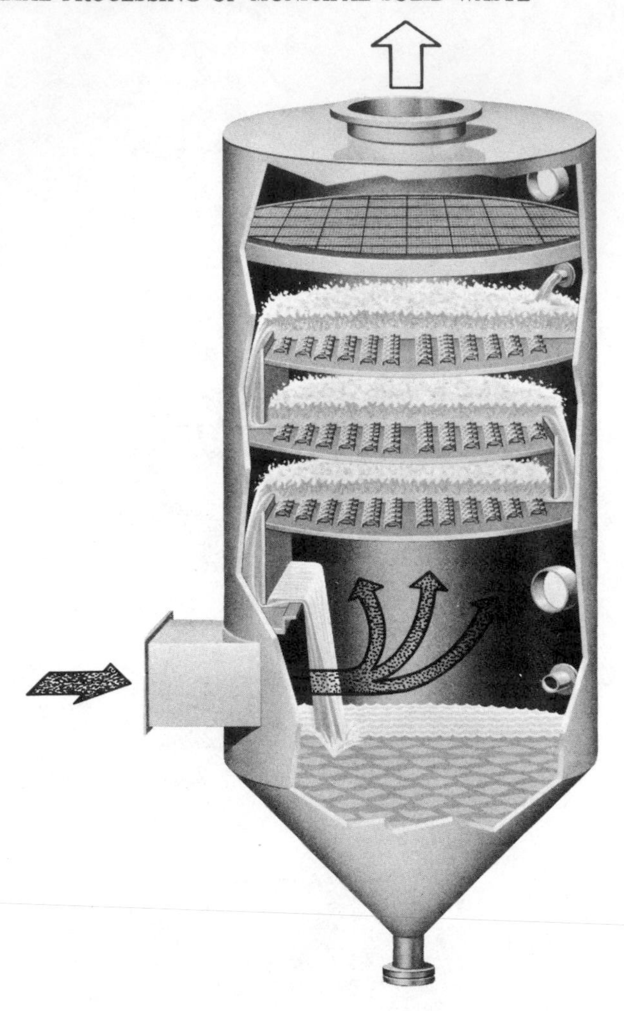

Figure 32. Gas-Motivated Orifice Type Scrubber (Courtesy of Koch Engineering).

In an ejector-type venturi, high-pressure liquid pumps rather than fans supply the motive power. As shown in Figure 33, water is supplied through a high-pressure spray nozzle to the venturi section of the scrubber. The gas and particulates are drawn into the scrubber and entrained by the liquid entering the venturi. Compression of the gas in the venturi creates the necessary pressure differential across the scrubber unit.

The gas and liquid droplets are intimately mixed by the turbulence created in the venturi, and the wetted particulates are separated in a section following the venturi using baffles or other devices.

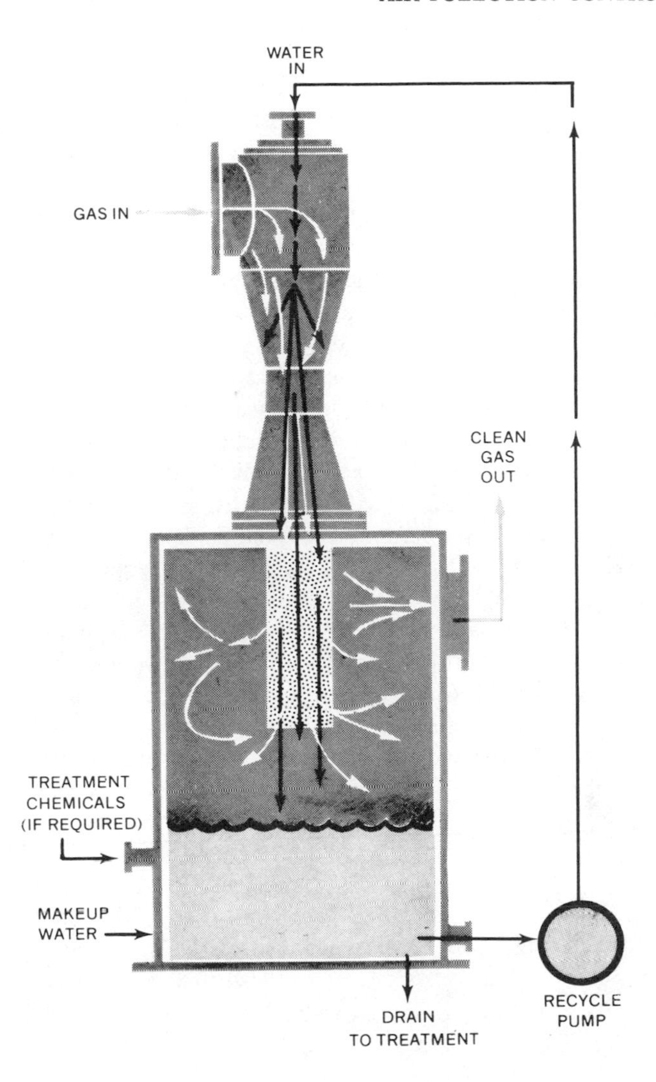

Figure 33. Liquid-motivated jet ejector scrubber (Courtesy of Croll-Reynolds).

The total water requirement is usually in the range of about 5.4 cubic meters per 100 cubic meters of flue gas (about 40-50 GPM/1000 CFM), primarily recycled water. Makeup water requirement is determined by the rate of solids removal from the gas. The water pressure required is typically near 5.8 atmospheres absolute (about 70 psig).

Energy requirements are similar to venturi scrubbers, but the need for a fan can be avoided where there are no other large draft requirements in the system. There are no known applications of ejector type venturi scrubbers to incinerators, but the possibility is worth investigating.

Since scrubber water requirements are high, recirculation is usually practiced, both to minimize makeup water and to minimize the amount of wastewater to be treated. The ratio of recycle to makeup water is determined by the quantity of particulates to be removed, by the tolerance of the scrubber design to the concentration of both soluble and insoluble materials in the water, which tend to build up with increased recycle, and the amount of water evaporated or otherwise lost.

Incinerator stack gases contain acidic gases which are soluble in water. These gases dissolve during scrubbing and cause the water to become acidic. As a result, even stainless steel scrubbers have been known to corrode away. Therefore, pH control by alkali addition must be practiced. This has two other important effects. First, undesirable acidic gases such as hydrogen chloride, hydrogen fluoride, and sulfur oxides are removed to some degree; and second, some carbon dioxide is removed, increasing alkali consumption. The latter may also have an important regulatory effect. Since emission standards are based on a 12% carbon dioxide content, the lower carbon dioxide content exiting from a scrubber can require an even lower actual particulate emission rate. The regulations are not clear on this matter.

A prediction of efficiency versus pressure drop (energy input) for a venturi scrubber applied to a municipal incinerator has been made.[3] Figure 34 presents this data, assuming an inlet particulate concentration of 1.25 grains per standard cubic foot dry (2.87 grams per standard cubic meter dry) corrected to 12% CO_2. The data predict that the federal standard of 0.08 grains/standard cubic foot (0.18 grams per standard cubic meter) can be reached, if at all, only at very high pressure drop.

However, several venturi scrubbers have been applied to incinerators, operating data for which are also shown on Figure 35. These data do not follow the predicted performance curve, but more nearly approximates the efficiency curve for a 1-μ particle or "fine particles" ($<5\ \mu$).[21] It would appear that a pressure drop of at least 22-32 mm Hg (12-17 in. H_2O) is required to achieve the federal standard. A "clear stack" (*e.g.*, 0.07 g/SCM or 0.03 gr/SCF) may require more than 37 mm Hg (20 in. H_2O), although the scrubber water vapor plume tends to reduce this requirement by masking the particulate opacity. The cleaned gases usually leave a scrubber at a temperature near 65.6°C (150°F), saturated with water vapor, thereby virtually always having a visible appearance from condensation of water vapor into fine droplets in the atmosphere. Commonly referred to as a steam or water vapor plume, this phenomenon is exempt from

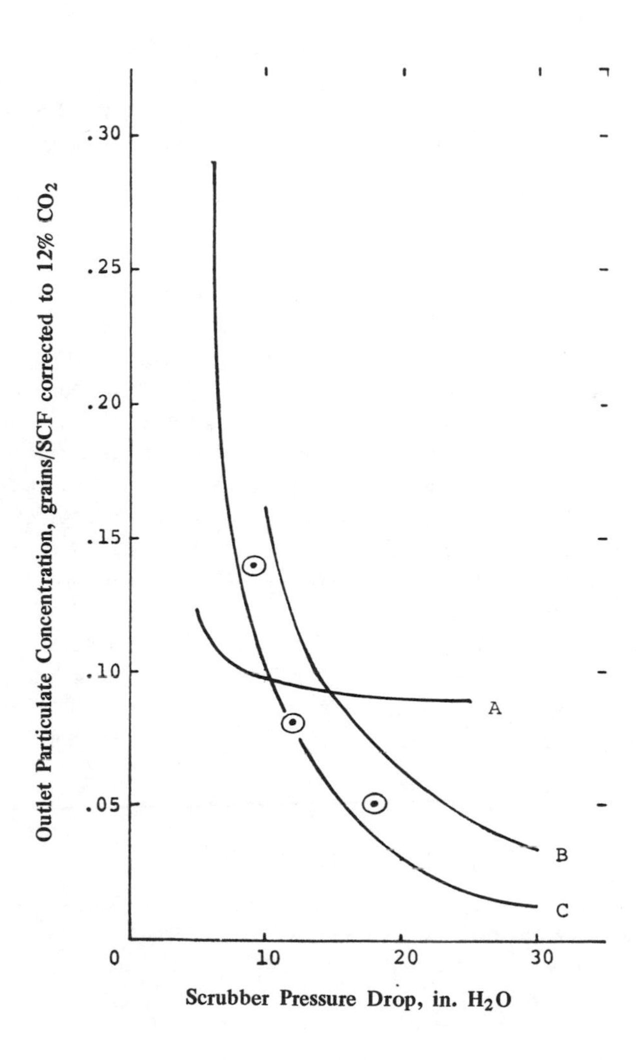

⊙ represents actual data from scrubbers on municipal incinerators

A. represents predicted performance of venturi scrubber on municipal incinerator[3]

B. represents predicted performance of venturi scrubber on fine particles (<5 μ)[21]

C. represents one manufacturer's claimed performance of venturi scrubber on 1-μ particles

Note: grams/SCM = 2.29 x grains/SCF mm Hg - 1.87 x in. H_2O

Figure 34. Performance curves for venturi scrubbers.

opacity regulations, but has other effects which may require control. Steam plume control is discussed elsewhere in this chapter.

Wastewater from scrubbers can often be used to quench the furnace residue prior to treatment or disposal, thereby reducing both water and treatment costs.

Fabric Filters

Fabric filters or baghouses are widely used in industrial applications, but only a few have been built for refuse incineration in the U.S. and Europe. In this device, the particulate-bearing gas stream is passed through a fabric filter medium of woven or felted cloth, which traps the particulates and allows the gas to pass through the pores of the fabric. These pores are as large as 100 microns, but even sub-micron particles are captured due to a buildup on the cloth of a fragile porous layer of collected particles which blocks the pores. For various economic and practical reasons, fabric filters are virtually always constructed in tubular form (bags) with numerous bags housed in a steel vessel (baghouse).

In order to operate continuously, the filter must be intermittently cleaned. This is accomplished by various means including manual, mechanical, or pneumatic shaking. The dislodged particulates fall to a hopper where they are removed by screw or other types of conveyor, as with precipitator hoppers. Figures 35 and 36 depict various baghouse designs.

Design parameters include:

- choice of fabric (based on gas temperature, humidity, and particle characteristics)
- size-length, diameter, and number of bags based on an empirically-obtained air flow to cloth ratio, and mechanical considerations.
- method of cleaning (based on particle characteristics and vendor preferences)
- method of precooling the gases to the operating temperature

There is no apparent reason why a fabric filter will not easily meet any existing particulate standard based on dry catch. The lack of significant use in incinerators may be due to various factors:

- dramatic sensitivity to temperature
- large space requirements
- difficult maintenance
- significant operating costs

The sensitivity of fabric filters to temperature is due to inability to withstand high temperatures (up to about 260°C or 500°F for the best fabrics); and low temperatures, where moisture adsorption and condensation will

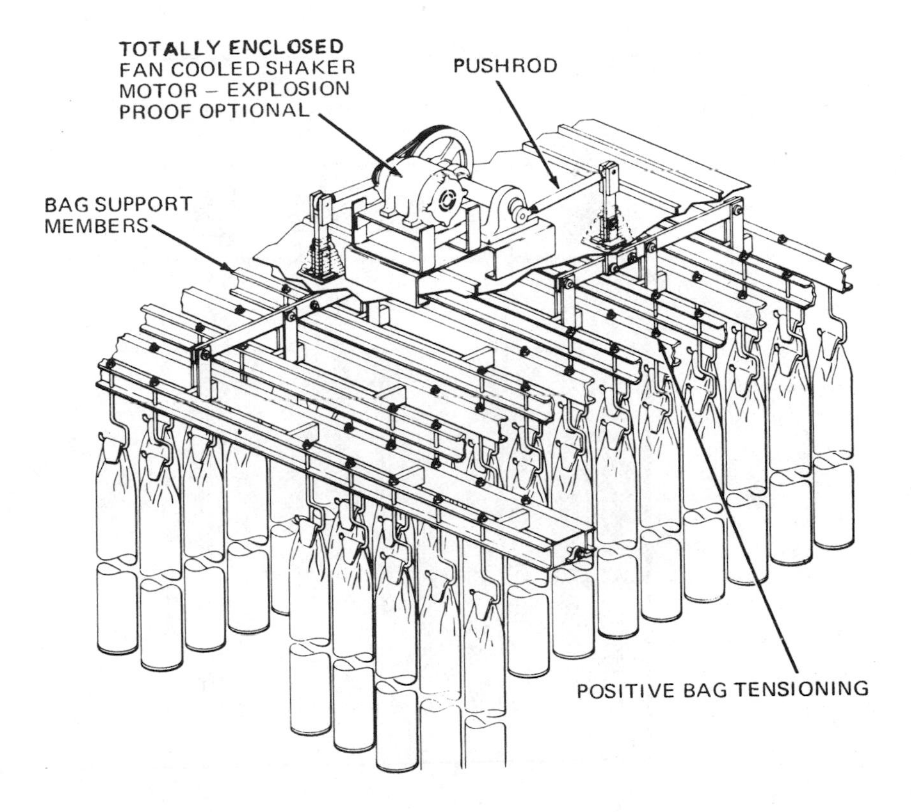

Figure 35. Filter baghouse with mechanical shaking (Courtesy of Buffalo Forge).

occur, blinding the bags and restricting flow. Operation consistently at temperatures near the fabric high-temperature limit may result in premature failure and frequent costly replacement, while even an occasional excursion above the temperature limit may cause failure or burnup of bags. Obviously, a well-designed, carefully-operated system to reduce and control flue gas temperature is an important part of any baghouse design.

Due to the large number of bags, baghouses require more space than scrubbers (excluding the wastewater system), but perhaps comparable space compared to precipitators. Maintenance is difficult because hundreds of bags are tightly spaced in a single housing. Physically finding and replacing broken bags is a dirty, difficult job. Complicated electronic and pneumatic timing and cycling devices for cleaning need specialized service.

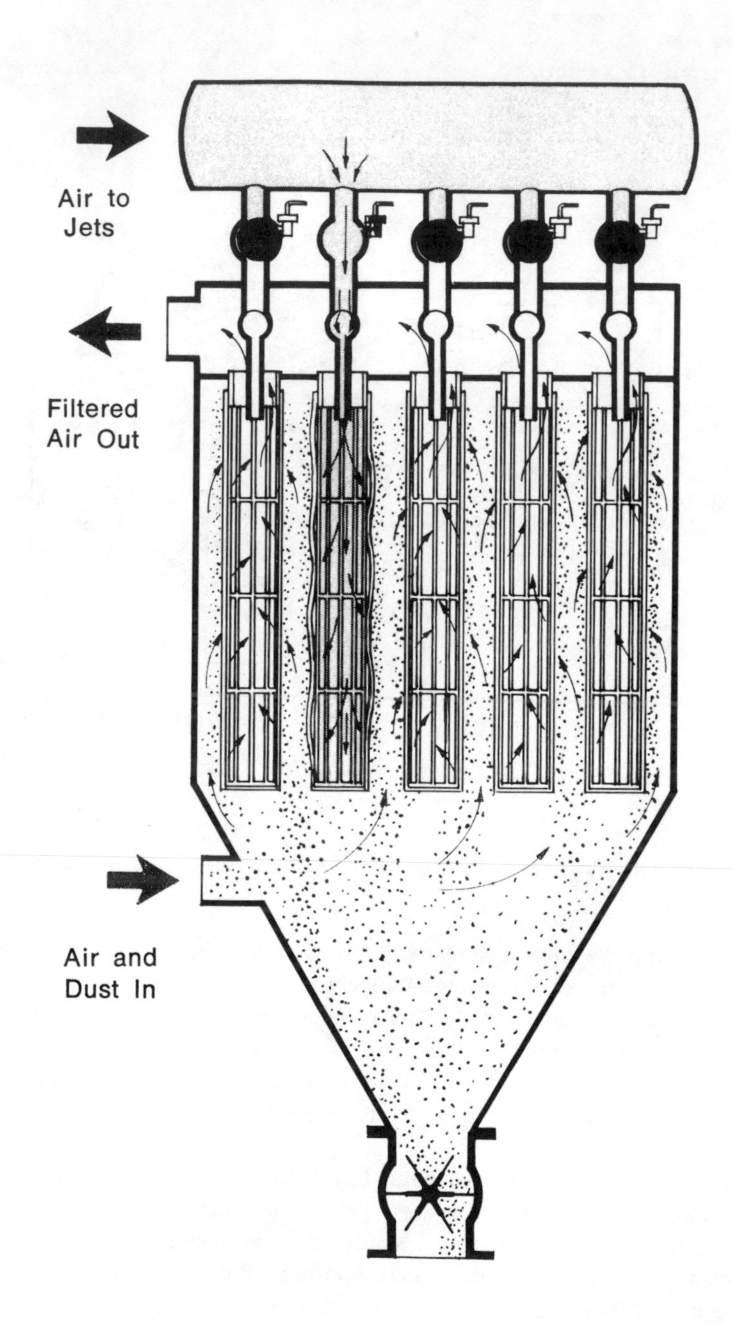

Figure 36. Filter baghouse with air pulse cleaning (Courtesy Carborundum).

High operating costs for baghouses are due both to the maintenance problem and to significant pressure drops across the bags. Operating costs are generally lower than for high-efficiency scrubbers, but greater than for electrostatic precipitators.

A pilot plant baghouse was operated with some success around 1959. One recent commercial installation on a municipal incinerator has apparently been operating reasonably successfully.[22] Pertinent data are presented in Table 62.

Table 62. Operating and Design Parameters for Fabric Filter Baghouse on Municipal Incinerator[22]

Air Flow, m^3/min (CFM)	5090 (180,000)
Air Temperature, °C (°F)	260 (500)
Fabric	glass fiber
Air/Cloth Ratio, m^3/min/m^2 (CFM/ft^2)	0.61 (2/1)
Bag Size	
diameter, m (in.)	0.14 (5.5)
length, m (ft)	4.27 (14)
Number of Bags (approx.)	4350
Method of Cleaning	reverse air
Design Pressure Drop, mm Hg (in. H_2O)	3.7-5.6 (2-3)

SELECTION OF PARTICULATE AIR POLLUTION CONTROL SYSTEMS

To choose between electrostatic precipitator, scrubber, or fabric filter systems for particulate removal from incinerator gases, parameters should be compared for each specific situation.

- Initial Cost–including cooling systems, fans, stacks, waste disposal, and other items dependent on the method of particulate control.
- Operating Costs–including power, water, maintenance, labor, and waste disposal.
- Reliability–considering best possible estimates for downtime; the effect of downtime on other operations, sensitivity to upsets and ranges of operating conditions; possible degradation of performance with age, and problems which are induced in associated equipment.
- Environmental and Other Considerations–including ability to meet and exceed emission standards, removal of nonparticulate pollutants, effect on the air quality of surrounding areas, the possibility of undesirable plumes, and the availability of facilities for waste disposal.

The initial cost for particulate control systems tends to be comparable when the complete system is considered, *e.g.,* including gas coolers, hoppers, and conveyors for precipitators; alloy metal construction, alkali addition, water supply, wastewater disposal, and water vapor plume control for scrubbers; gas coolers, hoppers, conveyors, and pulse air supply for baghouses. However, a definitive estimate is necessary where cost is a primary consideration.

Energy requirements probably represent the single most important difference between systems. Because of low pressure drops through an electrostatic precipitator, the total energy requirements are low, even though power is required for the corona discharge and the rappers and heaters. Fabric filter pressure drops are higher, requiring more energy, but scrubber energy requirement is by far the greatest of the three systems. Table 63 illustrates this difference. The importance of differences in energy requirements is at least partially dependent on the degree of energy recovery practiced in the incinerator and the availability of this energy for internal use.

Table 63. An Illustrative Comparison of Energy Requirements for Particulate Control Systems[a]

System	Gas-Motivated Scrubber	Fabric Filter	Electrostatic Precipitator
Gas Pressure Drop mm Hg (in. H_2O)	28.0 (15.0)	9.3 (5.0)	1.9 (1.0)
	KW per 1000 m^3/min (hp/1000 CFM)		
Fan Power	103.2 (3.92)	34.5 (1.31)	6.8 (0.26)
Pump Power	2.1 (0.08)	– –	– –
Electrostatic Power	– –	– –	15.8 (0.6)
Total Power[a]	105.3 (4.00)	34.5 (1.31)	22.6 (0.86)

[a]This table is based on a hypothetical calculation for approximately equivalent particulate removal efficiency. It does not necessarily include sufficient fan power for all furnace and duct pressure drops. Fan efficiency is approximately 60%. Power for heating hoppers (electrostatic precipitators, baghouses) or for tracing water lines (scrubbers) not included. These are variable depending upon design and location.

Since all of these systems are highly automated, operating labor requirements are essentially comparable and low when the systems are operating properly. Maintenance material and labor requirements for all of these particulate control systems can be very significant when designs and preventive maintenance are inadequate. Thus, differences between systems may be less important than the care which is tendered toward adequate design and operation.

Acid dew point corrosion of metal surfaces can be a problem in all cases, but it is more likely to occur when gases are cooled with spray water (rather than with steam boilers), such as is sometimes done ahead of either precipitators or baghouses and always an integral part of scrubber operation. Serious buildup of pressure drop and plugging can be serious problems with scrubbers or baghouses, but seldom occur with precipitators. Problems of hopper operation, which can occur with baghouses and precipitators, are obviously not a part of scrubber operations. Moderate excursions of temperature may have relatively minor effects on precipitator and scrubber operations, but can have drastic short- or long-term effects on filter bags, necessitating frequent changes which require significant labor and downtime.

The performance of all of the particulate control systems considered here can, at times, deteriorate. Dust buildup on either discharge or collection electrodes will cause diminished precipitator performance. Discharge electrodes in precipitators are subject to deterioration and breakage, sometimes shorting out a section of the precipitator and reducing its effectiveness. Hopper bridging can cause problems in both precipitators and baghouses. The development of breaks in filter bags can have drastic effects on baghouse performance. Deterioration in scrubbers can be caused by failure of spray nozzles, mist eliminators, poor pump performance, and other means.

Dry flyash disposal, as usually practiced with electrostatic precipitators and baghouses, is considered advantageous, but, unless the flyash is carefully handled, a considerable amount of fugitive emission can occur. The removal of solids in a slurry from scrubbers is less objectionable with incinerators than in other applications because this system can be integrated with the residue system for common water recycle and residue disposal facilities.

As noted earilier, the scrubber has at least one major advantage over dry methods for particulate removal. This advantage is the ability to simultaneously remove a significant portion of the gaseous emissions. However, present regulations do not require this control, and, if later necessary, it is possible to add efficient, low-energy gas scrubbers such as packed column or other types following the particulate control device.

To plan for this possibility, it is important to provide sufficient extra space and some static pressure allowance in the fans. The advantage of simultaneous gaseous emission control can also be met by using wet electrostatic precipitators as is done in certain other applications.

The major drawback to scrubbers, in addition to the high energy requirement, is the formation of visible moisture plumes with the possibility of icing and condensation. This problem is discussed in a subsequent section.

The foregoing discussion has been designed to show that each approach to particulate control has certain advantages and disadvantages. However, sufficient information is available to allow a logical selection based on careful cost and technical feasibility studies. Electrostatic precipitators have been chosen for most recent projects.

Combined Refuse/Fossil Fuel Firing

When firing prepared refuse with coal in an existing boiler, it will usually be necessary to adapt the available air pollution control equipment to the new mode of operation. In most coal-fired boilers particulate control is accomplished by high-efficiency electrostatic precipitators. It is expected that in many cases these precipitators can be used for combining firing with little or no modification. However, where relatively low-sulfur coal is being fired with exit flue gas temperatures greater than about 270-280°F, the addition of low-sulfur refuse may introduce, or make worse, resistivity problems due to the lack of sufficient sulfur trioxide conditioning agent naturally present from the coal sulfur. The little data available on this point strongly suggest this possibility.[22,23] In this situation, gas conditioning with sulfur trioxide or similar additives, or extensive precipitator modification might be necessary to meet emission regulations.

The firing of refuse with oil, which is being considered, may require installation of special particulate control equipment, if the boiler is not already so equipped. Some oil-fired boilers have been converted from coal and do have bottom ash removal, mechanical separators and/or electrostatic precipitators which could be useful, but careful analysis is required for each situation to assess potential air pollution control and other problems.

Emissions from the dust collection cyclones used with the hammermill and air classification system in a St. Louis demonstration plant have been measured.[24] These measurements indicate that careful design and operating practices will be necessary to avoid emissions from unit operations used to convert solid waste to fuel.

Pyrolysis

The pyrolysis plants which are being built or considered can be separated into two classes as far as air pollution control is concerned. First, there are those plants which produce primarily fuels for sale and use in other locations. These plants should not be prone to very serious air pollution control problems, except for those problems generally associated with handling municipal solid waste (such as odors), and for possible emissions from miscellaneous special unit processes such as shredding, air classifying, and drying. The second class of plants are those that internally burn the fuel produced to make steam for sale. In these plants, particulate control could be as difficult as in incinerators, with evaluation criteria similar to those already discussed.

Since pyrolysis processes available are primarily proprietary, the vendors have been specifying the entire plant, including air pollution control systems. This situation is expected to continue for some time. The limited information available on air pollution control for commercially available pyrolysis processes is covered in Chapter 3.

GASEOUS EMISSIONS

The overwhelming quantity of stack emissions from incinerators consists of relatively innocuous gaseous combustion products, namely carbon dioxide (CO_2) and water (H_2O), and unused oxygen (O_2) and inert nitrogen (N_2) from the combustion air. The relative and absolute quantities of each are determined primarily by

the composition of the refuse
the amount of excess air used (deliberately as well as through leakage)
the amount of air and/or water used for flue gas cooling and cleaning.

Figure 37 shows typical gas emissions as a function of excess air. Extensive data for a range of solid waste compositions, excess air, and gas cooling possibilities have been generated and reported,[25] but as explained in the reference cited, calculated gas flow rates are somewhat high and temperatures lower than would actually be expected. Table 64 shows typical gas compositions for two types of incinerator and air pollution control combinations. These gases are generally of little concern, except for water plume formation and insofar as certain air pollution standards are based on a specific gas composition, *e.g.*, 12 volume % CO_2. Emissions of various inorganic- and organic-contaminating gases are of concern and are considered in subsequent sections of this chapter.

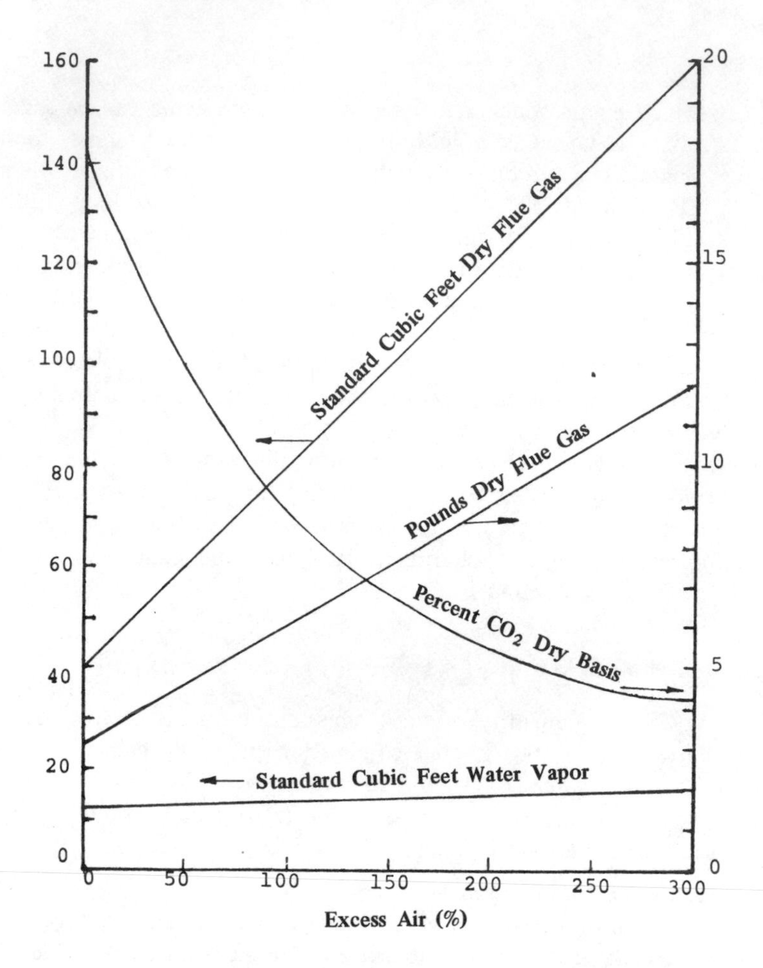

Figure 37. Gross products of combustion per pound of typical waste.[15]

Water Vapor Plume Control

High water vapor concentrations, under certain weather conditions, will produce a visible plume. Given the temperature and water concentration in the stack gas, the ambient temperature and humidity conditions under which the plume will form can be predicted.[28] Water vapor plumes are not regulated, but practical considerations, such as icing of nearby roads or buildings, fogging over roadways, or psychological reaction to a visible emission, may necessitate at least some degree of plume control.

Table 64. Typical Gas Compositions for Conventional and Steam-Generating Incinerators

Type of Incinerator	Waterwall[9]	Refractory[26,27]
Refuse Heating Value (HHV) cal/g (BTU/lb)	2420 (4360)	–
Refuse Firing Rate, MT/hr (ST/hr)	15.2 (16.7)	8.4 (9.2) (Rating)
Air Cooling/Air Cleaning Method	Boiler/Electrostatic Precipitator	Caustic Scrubber
Excess Air, %	71.7	180 (estimated)
Stack Exhaust		
Temperature, °C (°F)	211 (411)	57-77 (135-170)
Volume % CO_2, wet basis (dry basis)	9.1 (10.5)	3.7-3.2 (4.8)
Volume % O_2, wet basis (dry basis)	7.8 (9.0)	10.1-8.6 (13.0)
Volume % N_2, wet basis (dry basis)	69.8 (80.5)	64.2-54.2 (82.2)
Volume % H_2O, wet basis	13.3	22-34
Flow, CM/min @ temperature (ACFM @ temperature)	2400 @ 211°C (84,700 @ 411°F)	2119 @ 77°C (74,800 @ 170°F)

As noted earlier, visible stack exhaust plumes will be experienced when scrubbers are used, and may be experienced under certain weather conditions with other devices. These plumes are caused by condensation of water vapor into fine droplets when the hot, humid stack gases contact cold ambient air. Under some circumstances, it may be necessary or desirable to eliminate the visible plume.

The determining factors in plume formation are:

- stack gas temperature
- stack gas water concentration (humidity)
- ambient air temperature
- ambient air humidity

Since ambient air conditions obviously are not controllable, the only methods of water vapor plume control are to increase stack gas temperature and/or to reduce stack gas water concentration. Control methods may be designed for continuous or intermittent use, as desired.

Stack gas temperature can be increased by a stack burner which injects very hot combustion gases directly into the stack, by mixing with a warmer gas from another source, or by heat exchange with a source of heat (for example, steam or furnace flue gases). The first two approaches may also achieve a reduction in water concentration. Although relatively simple, hot gas addition is undesirable because of auxiliary fuel requirements. The increased gas flow also may require a larger diameter stack.

Heat exchange with the hot furnace flue gases is expensive because of the large amount of heat exchange surface required. However, if the heat exchange surface could be placed so as to remove heat directly from the combustion zone, it is theoretically possible to decrease excess air normally required for temperature control in conventional incinerators, thus increasing furnace capacity as limited by solid handling capability. At least one system has been built for heat exchange with hot flue gas, but none are known to operate by extraction of heat directly from the combustion zone. Unlike the systems discussed in the previous paragraph, indirect heat exchange does not decrease moisture content, which could even increase if excess air is reduced.

Plume control by reducing stack gas moisture content may be even less attractive than the methods described. All things considered, plume control should be avoided unless a very serious local environmental condition, such as road icing, results. In this case, intermittent control with the use of fuel for a brief duration may be the best choice. Such systems have been used on power plants.

Carbon Monoxide and Organic Gas Emissions

Poor combustion can result in emissions of carbon monoxide, hydrocarbons, oxygenated hydrocarbons, and other complex compounds. Some of these emissions are the source of odors which used to be associated with incinerator operation. While no data have appeared for modern facilities, improved combustion efficiency and control probably have reduced these emissions to insignificant proportions. Table 65 lists some measurements which have been made on older incinerators.

Handling and storage of solid waste, as well as furnace leakage, can create odors which may be detected both within the working areas of the facility and in nearby residential or commercial areas. No specific emission regulations exist which would control these emissions from incinerators. However, ambient air quality standards for carbon monoxide and hydrocarbons do exist (Table 66).

Inorganic Gas Emissions

Minor quantities of sulfur oxides, ammonia and halide gases are produced from the sulfur, nitrogen and halide (chlorine, bromine, fluorine)

Table 65. Carbon Monoxide and Organic Gas Emissions from Incinerators

	Concentration in Emitted Gas, ppm by volume			
			Refractory Incinerator[30]	
Gaseous Compound	Refractory[a] Incinerator[29]	Refractory Incinerator With Caustic Scrubber[26]	With Scrubber	Without Scrubber
Carbon Monoxide	<100-900	<25	<1000	<1000-3000
Hydrocarbons[b]	0.5-240	0	0	0
Oxygenated Hydrocarbons				
Aldehydes[c]	0.3-9.2	0.12	1-9	1-22
Organic Acids	0.1-1.0	1.6	–	–

[a]Probably with spray chamber.
[b]Reported as methane.
[c]Reported as formaldehyde.

Table 66. Federal Ambient Air Quality Standards for Gaseous Pollutants[12]

	Primary Standard	Secondary Standard
Carbon Monoxide	10 mg/m^3 (9 ppm) maximum 8-hr concentration[a]	same as primary
	40 mg/m^3 (35 ppm) maximum 1-hr concentration[a]	same as primary
Hydrocarbons	160 μg/m^3 (0.24 ppm) maximum 3-hr concentration (6-9 AM)[a]	same as primary
Photochemical Oxidants	160 μg/m^3 (0.08 ppm) maximum 1-hr concentration[a]	same as primary
Sulfur Oxides[b]	80 μg/m^3 (0.03 ppm) annual arithmetic mean	1300 μg/m^3 (0.5 ppm) maximum 3-hr concentration
	365 μg/m^3 (0.14 ppm) maximum 24-hr concentration[a]	
Nitrogen Dioxide	100 μg/m^3 (0.05 ppm) annual arithmetic mean	same as primary

[a]Not to be exceeded more than once per year.
[b]Measured as sulfur dioxide.

content of the solid waste. Nitrogen oxides emitted may result from the nitrogen content of the waste or high-temperature oxidation of nitrogen in the air. All these emissions are expected to be directionally, but not necessarily quantitatively, related to the concentration of the source element in the waste burned. Table 67 shows typical quantities emitted.

Hydrogen chloride (HCl) has been of particular concern because of increasing emissions due to increased disposal of polyvinyl chloride (PVC) and other halide-containing plastics and aerosols, because of possible health effects and because of the possibility of corrosion, especially of tube metal surfaces in steam-generating systems (discussed in Chapter 2). Table 67 shows an approximate correlation between the PVC content of the refuse and HCl emissions. It is believed that almost all of the chlorine in PVC (greater than 50% chlorine) is converted to HCl but that some of the HCl reacts with particulate matter and is removed by particulate control equipment.

Hydrogen chloride and fluoride, sulfur oxides, nitrogen oxides, and some oxygenated hydrocarbons are acidic with at least some solubility in water. Therefore, when flue gas is exposed to liquid water, such as in a quench chamber or scrubber, or even due to cold wall condensation, very corrosive conditions exist which will attack even stainless steel. Neutralization, with caustic or other alkaline materials, both enhances removal and helps prevent acid attack in wet systems.

There are no specific federal emission standards for gaseous emissions from incinerators, but ambient air quality standards do exist for sulfur oxides, and nitrogen dioxide (Table 66). Though nitrogen oxide and sulfur oxide emissions from power plants are limited by some regulatory standards, no such limitations are known to exist for incinerators.

As shown in Table 67, ammonia is sometimes reported as a constituent of incinerator effluent gases. The ammonia may result from refuse decomposition reactions. Wet pollution control systems would be expected to remove at least some ammonia, especially where the water is acid or near neutral. No specific regulations exist for ammonia control. Although information is only minimal as to the extent of ammonia emissions, it does not appear to be a serious problem.

Control of Gaseous Emissions

Carbon monoxide and hydrocarbon emissions, including odorous compounds, are effectively controlled by a well-designed combustion chamber and careful control of operating conditions. Odors from waste handling and storage can be controlled by drawing the combustion air into the plant through these areas so that the odor-bearing gases will be burned

Table 67. Sulfur Oxide, Ammonia, Nitrogen Oxide, and Halide Gaseous Emissions from Incinerators

Gaseous Compound	Concentration in Emitted Gas, ppm by volume						
	Refractory Incinerator[30]			Refractory Incinerator	Refractory Incinerator[31] At Furnace Exit		
	With Scrubber	Without Scrubber	Refractory[a] Incinerators[29]	with Caustic Scrubber[26]	"Normal Refuse"	2% PVC Added	4% PVC Added
Nitrogen Oxides	22-58	58-92		2.6	23-25	40.6	46.2
Sulfur Oxides	–	–	54-109[b]	14.6[c]	33-40	37.6	38.6
Ammonia	–	–	–	28.6	–	–	–
Chlorine	–	–	–	1.9		none detected	
Hydrogen Chloride	–	–	38-113	11.3	455-732	1990	3030
Hydrogen Fluoride	–	–	0.6-0.9	–	0.9-2.3	4.6	2.8

[a]Probably with spray chamber.

[b]$SO_2 + H_2SO_4$.

[c]SO_2 only.

in the incinerator. Similarly, leakage from the furnaces can be prevented by maintaining a slight negative pressure within.

Inorganic gaseous emissions do exist in relatively low concentrations, but they need not be controlled in the absence of specific regulations. Where water scrubbers are used for particulate control, significant removal of hydrogen chloride, sulfur dioxide, and nitrogen dioxide (but not nitric oxide) will occur. Where only electrostatic precipitators or baghouses are used for particulate control, only a small amount of these gases will be removed, limited to that quantity which reacts with or adsorbs onto particulates.

MONITORING

Although there are no specific requirements for air pollution monitoring, the following stack monitors may be useful for operational as well as environmental purposes and are often used.

opacity (smoke density)
oxygen
carbon dioxide
combustibles
television

Opacity meters measure the transmission of light through the stack gases. They can be used as an indicator of whether particulate emission standards are being met, as well as to detect periods of poor combustion, characterized by "smoke-laden" flue gas.

Oxygen and/or carbon dioxide monitors measure these chemical constituents in the flue gas. They are useful as indicators of the amount of excess air being used, although usually only one or the other is provided. Combustible gas monitors measure organic content, and can be also used as an indicator of furnace combustion conditions. Television cameras are useful for monitoring smoke, water vapor plumes, and general surveillance.

In order to insure successful operation of monitoring instruments, provision for and practice of regular care, cleaning, calibration, and other maintenance are essential.

STACKS

Traditionally, stacks have been used to provide natural drafts for the furnaces as well as to disperse the "noxious" gases from previously uncontrolled incinerators. With the use of fans in modern facilities, the draft aspects of the stacks are less important. However, even with modern air pollution control devices, stacks are generally still required to disperse residual pollutants in order to avoid high ground level concentrations. Sophisticated dispersion modeling techniques, usually computerized, can predict air quality resulting from utilizing various stack heights and diameters, and can even aid in stack location to avoid "downdraft" effects due to buildings and hills. As discussed earlier, the choice of the number of stacks can also be affected by opacity considerations.

If water vapor plumes are experienced, stacks also function to disperse them before impinging on surfaces where icing or condensation can be harmful.

Numerous stack designs and materials, ranging from ordinary steel to corrosion-resistant steels (*e.g.*, Corten, stainless) to brick and mortar have been used, all with reasonable performance and expected lives. Stacks with hermetically-sealed inner liners to prevent condensation and chemical attack have been used for those installations requiring especially long-lived stacks.

REFERENCES

1. Environmental Protection Agency Standards of Performance for New Stationary Sources. 36 FR 24880ff. December 23, 1971.
2. New Jersey Department of Environmental Protection. Regulations on Incinerators; New Jersey Administrative Code, Chapter 27, Bureau of Air Pollution Control Subchapter 11, Incinerators; NJAC 7:27-11.
3. Niessen, W. R. *et al. Systems Study of Air Pollution From Municipal Incineration.* Volume I. Arthur D. Little, Incorporated. Cambridge, Massachusetts. U.S. Department of Health, Education and Welfare. National Air Pollution Control Administration Contract No. CPA-22-69-23. NTIS Report PB 192 378. Springfield, Va. March 1970.
4. Duprey, R. L. Compilation of Air Pollutant Emission Factors. Public Health Service Publication No. 999 AP-42. 1968.
5. Fernandes, J. H. Incinerator Air Pollution Control Proceedings, 1968 National Incinerator Conference. New York. May 5-8, 1968. American Society of Mechanical Engineers. pp. 101-116.
6. Jens, W. and F. R. Rehm. Municipal Incineration and Air Pollution Control. Proceedings, 1966 National Incinerator Conference. New York. 1966. American Society of Mechanical Engineers. pp. 74-83.
7. Chass, R. L. and A. H. Rose. Discharge From Municipal Incinerators. *Air Repair* 3(2): 119-22. November 1953.
8. Walker, A. B. Electrostatic Fly Ash Precipitation For Municipal Incinerators, A Pilot Plant Study. Proceedings, 1964 National Incinerator Conference. New York. May 18-20, 1964. American Society of Mechanical Engineers. pp. 13-19.
9. Stabenow, G. Performance of the New Chicago Northwest Incinerator. Proceedings, 1972 National Incinerator Conference. New

York. June 4-7, 1972. American Society of Mechanical Engineers. pp. 178-194.
10. Kaiser, E. R. Refuse Compositions and Flue Gas Analyses From Municipal Incinerators. Proceedings, 1964 National Incinerator Conference. New York. May 18-20, 1964. American Society of Mechanical Engineers. pp. 35-51.
11. Walker, A. B. and F. W. Schmitz. Characteristics of Furnace Emissions From Large Mechanically-Stoked Municipal Incinerators. Proceedings, 1966 National Incinerator Conference. New York. May 1-4, 1966. American Society of Mechanical Engineers. pp. 64-73.
12. Environmental Protection Agency Regulations on National Primary and Secondary Ambient Air Quality Standards. 40 CFR 50; 36 FR 22384. November 25, 1971. As amended by 38 FR 25678. September 14, 1973.
13. Ensor, D. S. and M. J. Pilat. Calculation of Smoke Plume Opacity From Particulate Air Pollutant Properties. *Journal of Air Pollution Control Association.* **21**(8): 496-501. August 1971.
14. Bump, R. L. The Use of Electrostatic Precipitators On Municipal Incinerators. *Journal of Air Pollution Control Association.* **18**(12): 807-809. December 1968.
15. De Marco, J. *et al.* Municipal-Scale Incinerator Design and Operation. PHS Publication No. 2012. U.S. Government Printing Office, Washington, D. C. 1969. (formerly Incinerator Guidelines - 1969), p. 53.
16. Ross, R. D. ed. *Air Pollution and Industry.* Van Nostrand Reinhold. New York. 1972. 489 pages.
17. Fife, J. W. Techniques for Air Pollution Control in Municipal Incineration. American Institute of Chemical Engineers Symposium Series 70(137): 465-473. 1974.
18. White, H. J. Resistivity Problems in Electrostatic Precipitation. *Journal of Air Pollution Control Association.* **24**(4): 313-338. April 1974.
19. Manual of Disposal of Refinery Wastes. Chapter 12. Electrostatic Precipitators. American Petroleum Institute Publication No. 931. Washington, D. C. June 1974.
20. Hall, H. J. Design and Application of High Voltage Power Supplies in Electrostatic Precipitation. *Journal of Air Pollution Control Association.* **25**(2): 132-138. February 1975.
21. Hesketh, H. E. Fine Particle Collection Efficiency Related to Pressure Drop, Scrubbant and Particle Properties, and Contact Mechanism, *Journal of Air Pollution Control Association.* **24**(10): 939-942. October 1974.
22. Bergmann, L. New Fabrics and Their Potential Application. *Journal of Air Pollution Control Association.* **24**(12): 1187-1192. December 1974.
23. St. Louis/Union Electric Refuse Firing Demonstration Air Pollution Test Report. EPA-650/2-74-073. U.S. Environmental Protection Agency. Washington, D. C. August 1974.
24. Shannon, L. J. *et al.* St. Louis Refuse Processing Plant: Equipment, Facility, and Environmental Evaluations. EPA-650/2-75-044. U.S. Environmental Protection Agency. Washington, D.C. May 1975. 122 pages (NTIS No. PB-243 634).

25. Technical-Economic Study of Solid Waste Disposal Needs and Practices. Combustion Engineering, Inc. Windsor, Ct. Report SW-7c. U.S. Department of Health, Education and Welfare. Bureau of Solid Waste Management. 1969. Volume IV.
26. Gilardi, E. F. and H. F. Schiff. Comparative Results of Sampling Procedures Used During Testing of Prototype Air Pollution Control Devices at New York City Municipal Incinerators. Proceedings, 1972 National Incinerator Conference. New York. June 4-7, 1972. American Society of Mechanical Engineers. pp. 102-110.
27. Ellison, W. Control of Air and Water Pollution From Municipal Incinerators With the Wet-Approach Venturi Scrubber. Proceedings, 1970 National Incinerator Conference. Cincinnati. May 17-20, 1970. American Society of Mechanical Engineers. pp. 157-166.
28. Rohr, F. W. Suppression of the Steam Plume From Incinerator Stacks. Proceedings, 1968 National Incinerator Conference. New York. May 5-8, 1968. American Society of Mechanical Engineers. pp. 216-224.
29. Carotti, A. A. and R. A. Smith. Gaseous Emissions From Municipal Incinerators. U.S. Environmental Protection Agency. Publication SW-18c. 1974. 61 pp.
30. Corey, R. C. *Principles and Practices of Incineration.* Wiley Interscience. New York. 1969. p. 82.
31. Kaiser, E. R. and A. A. Carotti. Municipal Incineration of Refuse with 2% and 4% Additions of Four Plastics. A Report to the Society of the Plastics Industry. New York. June 30, 1971. Table 12.

CHAPTER 7

LIQUID AND SOLID EFFLUENTS AND THEIR CONTROL

Even the most efficient thermal processing of solid wastes results in undesirable effluents. Incineration effluents are primarily inorganic, in the form of particulate and other chemical emissions in gaseous effluents, dissolved and suspended materials in aqueous effluents, and solid residues. As will be discussed in a subsequent section of this chapter, pyrolysis effluents may also be high in organic materials.

The unit processes which are commonly used in incinerator plant design and the associated discharges to air, land and water are shown in Figure 38. This chapter describes discharges to land and water, while air emissions were treated in Chapter 6 and recovery of useful materials from residues was covered in Chapter 4.

HANDLING AND STORING EFFLUENTS

Dumping, handling, and storage of municipal solid wastes produce dust and litter as well as odors. Therefore, the storage pits and tipping floor are areas which deserve major attention, if problems from these sources are to be averted.

Litter, consisting mainly of paper and other debris, results from accidental spillage in and around the plant. Dumping of refuse produces airborne dust which may be either organic or inorganic. Obnoxious odors result primarily from the putrefaction of food wastes and other organic materials. Odor problems are especially troublesome when waste is held in the storage pits for long periods. The combined effect of dust and odors, if uncontrolled, creates a condition which is very unpleasant for employees who work in these areas. Dust can also adversely affect instrumentation and controls, and mechanical and electrical equipment.

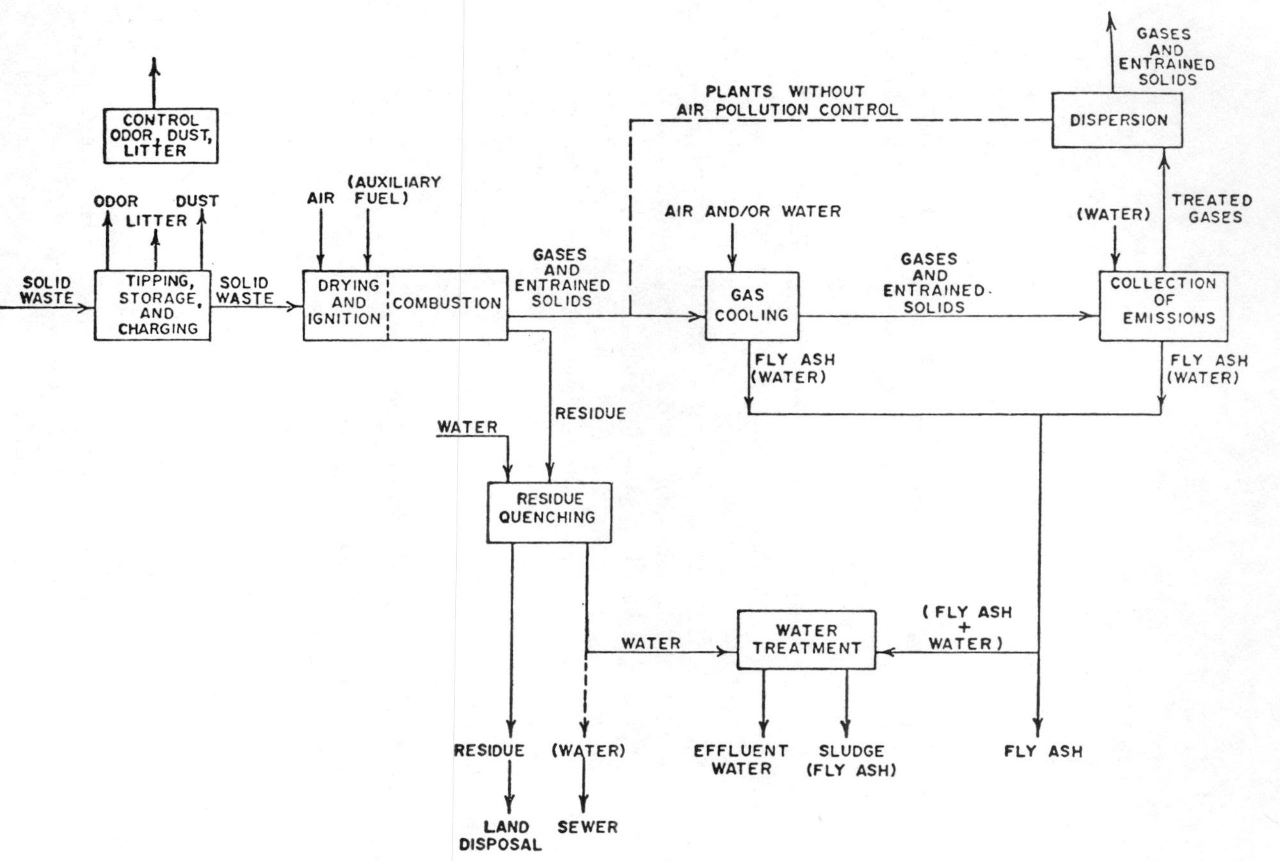

Figure 38. Diagram of the inplant systems based upon dry fly ash collection and conveying from cooling and collection operations. Alternatives for wet collection and conveying shown in parentheses.[1]

Litter is primarily a nuisance which creates an untidy appearance of the plant and grounds.

Frequent sweeping of the tipping floor effectively removes litter. Cleaning of the storage pit is facilitated if the pit is divided into sections. Each section of the pit should be emptied at frequent intervals so that putrescibles may be removed. The tipping floor and pit floor should be washed with cleaning/disinfecting solutions for control of odors and insects.

Wetting the solid waste in the storage pit by use of water sprays is the most frequently used means of dust control. Some incinerator plants control dust and odors in the pit by locating intake ducts for the forced draft fans within the pit areas, sweeping these pollutants into the fan intake and then into the furnace, thus preventing their dispersion throughout the plant.

The future may bring greater emphasis to the design of enclosed material handling systems to permit positive exhaust of air through purification systems. When such provisions are made, dust-laden air can be processed in bag filters or other means to prevent release to the atmosphere. Odors in exhaust air can sometimes be eliminated by addition of ozone[2] or by passing the air through beds of activated carbon.[3]

Provision should be, but is not always, made for treatment of water used for dust control, and for washdown of the tipping floor, charging floor, and storage pit. Runoffs of this type are frequently drained directly to sanitary sewers or surface waters. Preferably, all these wastewaters should be treated with the process wastewater.

INCINERATOR RESIDUES

Incinerator residues are defined here as the solid materials remaining after combustion. Classifications of residue include grate residue, grate siftings, and flyash. The quantity of residue produced, when front end resource recovery is not practiced, generally ranges from 20 to 35% by weight of the original refuse, but usually only about 5 to 15% by volume. The proportions of grate residue, siftings, and flyash depend to a large extent on the design of the incinerator and of the air pollution control equipment.

Grate Residue and Siftings

Residue discharged from burning grates consists of ash, clinkers, cans, glass, rocks, and unburned organic substances. Grate siftings are similar materials which have become sufficiently reduced in size to filter through the grate openings or to drop between the grate and the furnace wall.

Bulk densities of grate residue and siftings, as measured in one test, were 640 and 1055 kilograms per cubic meter (1040 and 1780 lb/cubic yard), respectively.[4]

Composition of combined incinerator residues from three sources are given in Table 68.[4] Since seawater was used for sluicing at the Oceanside plant, a portion of the residue measured was contained in the sluice water runoff. The Oceanside and Stamford plants employed rocking grates with comparatively large grate openings. The Washington, D.C. data were obtained from five batch-fed plants.

Table 68. Comparison of Residue Compositions[4]

	Oceanside Wt %	Stamford Wt %	Washington, D.C. Wt %
Metals and Mill Scale	19.85	23.58	29.5
Glass	9.48	36.63	44.1
Ceramics, Stones	1.51	4.73	2.0
Clinkers	24.11	17.23	–
Ash	16.10	14.08	15.4
Organic[a]	1.89	3.75	9.0
Residue Solids in Conveyor Runoff Water	27.06	0	–
	100.0	100.0	100.0

[a]Good measure of burnout.

Siftings may be recovered wet or dry depending on the incinerator design. However, most grate residues are recovered from quench water. The wet residues are drained as they are conveyed from the furnace. The siftings and wet residue are usually trucked to nearby landfill sites. These should be, but are not always, water-tight trucks. If resource recovery is practiced (for example, magnetic recovery of ferrous metals), the separation equipment is built into the incinerator plant.

Flyash

As explained previously, flyash is that portion of the solid residue from combustion carried by the combustion gases. It arises as a solid effluent after recovery from flue gases using various air pollution control devices. The flyash consists of dust, cinders, soot, charred paper, and other partially-burned materials. Most flyash particles range in size from 120 to less than 2 microns.[5] Size distribution within this range is extremely variable.

Flyash from efficient incineration is predominantly inorganic and consists largely of the oxides or salts of silicon, aluminum, calcium, magnesium, iron and sodium.[4] Compounds of titanium, barium, zinc, potassium, phosphorus and sulfur[4] may be present in small amounts. Trace quantities of many other elements may also be present.

Dry flyash is difficult to handle and can be easily picked up and scattered by the wind. Therefore, it should be stored in closed containers. If open storage is necessary, barriers should be erected and the surface of the ash pile kept moist with water sprays. Transportation to the final disposal site in covered trucks or closed containers is recommended. Some plants reduce dust problems by intermixing ash with wet residue, or by topping off the truck with a layer of wet residue.

Land Disposal of Residue

Although incinerator residue is comprised mainly of insoluble inorganic material, the small fractions of soluble inorganics and organics require land disposal methods usually classified as sanitary landfilling. Guidelines for sanitary landfill site selection, design, and operation are available.[6,7]

Sanitary landfill practices designed to avoid pollution of surface and ground waters, odors, rodents, insects and other vectors require spreading the solid wastes in thin layers, compacting to the smallest practical volume, and applying a compacting cover material at the end of each operating day, or more often. Practices for a landfill disposing of only incinerator residue may differ from disposing of mixed municipal solid waste because of the lower organic content (causing less gas formation) and higher density (requiring less compaction). However, measures designed to avoid water pollution–for example, the use of impervious membranes as a barrier against groundwater intrusion and leaching from rainfall, interception of rainfall and surface waters, and/or treatment of leachate–will most likely be similar for an incinerator residue landfill.

The amount of landfill leaching which will occur is dependent upon the composition of the residue, its permeability, and the degree of fusion of external and internal surfaces. When water contacts the residue, it almost invariably picks up fine solid particles which contribute to increased levels of suspended solids, turbidity, and BOD (from organic content) in the water. Because the presence of organics in the residue may lead to particularly harmful environmental effects, the degree of burnout during incineration is an important variable.[8]

There is a good deal of disagreement regarding the efficacy of landfill for disposal of incinerator residues. It has been determined that the water-soluble portion of the residue amounts to approximately 4.75 to 5.75% of the dry weight of material placed.[9] Data are generally not available, however, on the extent to which this material is removed by leaching, or the rate at which leaching takes place. Leaching tests conducted in Germany[10] suggest a relatively low mineral content of water after the elutriation of finely-ground incinerator residue with distilled water. The initial eluate contained approximately 115 mg/l minerals, and the tenth eluate contained less than 20 mg/l. By comparison, similar tests conducted using composted refuse produced concentrations of calcium and magnesium (as CaO and MgO) which were an order of magnitude higher than in the residue leachate. The explanation proposed was that as the residue reaches a combustion temperature of 800°C, the salts are converted to oxides which are insoluble, or only slightly soluble. Also, it was noted that glassy insoluble substances are formed with the silica present (*e.g.*, calcium silicate and magnesium silicate). It appears that additional testing is needed to determine the extent of the leaching problem and to develop control methods.

INCINERATOR WASTEWATER

The process wastewater from incinerator plants is contaminated by both dissolved and suspended materials. To prevent pollution of streams and underground water, some form of treatment is usually required prior to discharge.

It is important to distinguish between the treatment required prior to discharging to a sewer system which sends its water to a municipal treatment plant, and discharging directly to the environment, for example, a marsh, river, or tidal basin. In the former instance, minimal treatment such as settling and possibly pH adjustment may be adequate. In the latter, the treatment system must be designed to remove the objectionable contaminants to a level consistent with federal, state and local water quality discharge regulations.

Process Wastewater Sources and Quantity

The sources of process water from an incinerator plant include feed chute water jackets, furnace wall cooling, residue quenching, residue and flyash conveying, wet scrubbers, wet baffles, wet bottoms, and settling chambers. Intermittent uses include storage pit sprays, and floor and pit washings.

The quantity of water used in incinerator plants varies widely, depending upon the extent and mode of water uses in air pollution control equipment, residue conveying,and stack gas temperature control. An individual estimate must be made for each incinerator design, but, based on previous estimates,[11,12] the quantity of water discharged can exceed 12 tons per ton of solid waste processed (2900 gal per 2000 lb) when scrubbers are used. This quantity may be cut by a factor of two in the absence of scrubbers, and to much smaller values when extensive recycle is used, for example, as little as 2-3 tons per ton of waste.[12]

Other Incinerator Wastewaters

Other than the process wastewater, an incinerator plant will discharge the usual sanitary wastes and runoff waters. The sanitary wastes are usually discharged to a sanitary sewer for treatment in the municipal sewage plant. The quantity can be estimated from the number of employees.

Runoff water varies with washdown procedures, precipitation, terrain, and soil characteristics. The necessity to deal with surface and runoff waters in residue landfills was discussed in the "Incinerator Residue" section. Although runoff water is less a problem where only the incinerator or other thermal processing unit has to be considered, it is nevertheless a real one. Water can be contaminated by the litter and dust which invariably is associated with solid waste handling.

A preferred solution to the problem is to direct storm sewer effluents from handling areas to the process water system, but regulations do not always permit this approach if the process water is discharged to a municipal treatment plant. In that case, on-site treatment might be required. Uncontaminated runoff waters can be handled in storm sewers.

Wastewater Quality

The quality of water discharged from the various process units also varies widely from one plant to another, and daily variations occur within the same plant. Variations result from nonuniformity of solid waste composition and changes in water usage. Some important wastewater characteristics are:

- Temperature
- Dissolved Oxygen (DO)
- Biochemical Oxygen Demand (BOD)
- Chemical Oxygen Demand (COD)
- Hydrogen Ion Concentration (pH)
- Alkalinity
- Hardness
- Total Solids
- Total Dissolved Solids
- Suspended Solids
- Settleable Solids
- Phosphates
- Nitrates
- Chlorides
- Fluorides
- Heavy Metals
- Odor
- Color

Water analyses for several incinerator plants, including quench water, scrubber water and final effluent water, are given in Table 69. The extent of variation in the amounts of each contaminant is apparent.

Table 69. Typical Wastewater Analyses[11]

	Quench Range	Scrubber Range	Final Effluent	
			Range	Avg.
pH	3.9-11.5	1.8-9.4	4.5-9.9	–
Temperature, °C	20-54	28-74	18-52	32
(°F)	(68-130)	(82-165)	(65-125)	(90)
Suspended Solids, mg/l	140-1860	90-1350	40-580[a]	210
Dissolved Solids, mg/l	360-2660	520-8840	320-4060	1190
Total Solids, mg/l	610-3960	610-9160	610-4200	1400
Alkalinity, mg/l $CaCO_3$	90-720	0-80	15-310	135
Chlorides, mg/l	98-850	180-3540	95-1710	455
Hardness, mg/l	95-980	190-3430	100-480	240
Phosphates, mg/l	0.5-58	3-90	1-67	14

[a]After settling.

Residue quench water, which may be either basic or acidic, contains moderate concentrations of dissolved minerals and often high concentrations of suspended solids. Temperatures are not exceptionally high, ranging from 20-54°C (68-130°F), but cooling is sometimes required.

Acidic conditions usually prevail in scrubber water. Dissolved mineral content is much higher than in quench water due to combined effects of low pH and the practice of recycling a portion of this water back to the scrubber. Average values for suspended solids in the scrubber water are lower than those presented for quench water. This may be related to the design of the scrubber, which permits settling prior to reuse of the water. The temperature of scrubber water is considerably higher than for quench water and ranges from 28-74°C (82-165°F).

Combined wastewater from all sources may be either acidic or basic in nature as shown in Table 69, which indicates a pH range of from 4.5 to 9.9. Dissolved solids vary from 320 to 4060 mg/l, presumably depending upon the extent to which water is recycled. Suspended solids content ranges from 40 to 580 mg/l, the lower values representing water sampled after settling has taken place. Temperatures ranged from 18-52°C (65-125°F), the higher values being of concern from the standpoint of thermal pollution and dissolved oxygen depletion of the receiving water.

Data in Table 70 for cooling-expansion chamber spray water show trace contaminants such as fluoride, iron and ABS (alkyl benzene sulfonate).[13] It should be noted that the fluoride value of 7.8 mg/l is significantly higher than the permissible level for public water supply

Table 70. Chemical Quality[a] of Cooling-Expansion Chamber Water Discharge at Incinerator No. 1[13]

Constituent		Test 1	Test 2	Test 3	Test 4
pH		7.9	5.5	3.5	6.2
Alkalinity	($CaCO_3$)	35.0	9.0	0	10.0
Nitrate	(NO_3)	1.5	1.90	2.00	2.00
Phosphate	(PO_4)	0.2	0.61	0.39	0.17
Chloride	(Cl)	582.0	453.0	567.0	422.0
Fluoride	(F)	4.5	6.40	4.40	7.80
Calcium	(Ca)	330.0	220.0	255.0	250.0
Sulfate	(SO_4)	338.0	238.0	188.0	300.0
Sodium	(Na)	85.0	63.0	73.0	60.0
Potassium	(K)	27.0	14.8	16.7	14.0
Iron	(Fe)	1.6	0.93	5.77	0.50
ABS	(ABS)	1.91	0.1	0.19	0.21

[a]All values are expressed in mg/l except pH.

(1.2 mg/l). Other data for minor contaminants in scrubber water are shown in Table 71. These include cyanide, phenols, iron, chromium, lead, copper, zinc, manganese, aluminum, and barium. Of these, all but total chromium exceeded the Florida quality standards for incinerator effluents. Additional data in Table 72[14] show the effect of pH in fly-ash water on the concentration of various metal ions found in the water sampled. In almost every instance, the water having the higher acidity (lower pH) contained larger amounts of metals.

Data on oxygen demand characteristics of incinerator water are seldom reported in the literature. One source[15] reports 5-day BOD determinations (expressed in mg/l) for water from the following sources: residue conveyors - 618, 750, 560, 605 (four different incinerator plants); ash hopper - 700; flyash disposal - 3.2; and lagoon - 54. BOD of the combined wastewater from the incineration of municipal solid waste would be expected to be similar to that of domestic sewage, averaging about 200 to 300 mg/l.

Table 71. Scrubber Water Chemical Characteristics[13]

Constituent		Quality Standard[a]	Raw Water	Scrubber Effluent
Iron	(Fe)	0.3	0.35	1.65
Cyanide	(Cn)	0	0.21	5.19
Total Chromium	(Cr)	1.0	0.0	0.13
Lead	(Pb)	0.50	0.0	1.30
Phenols		0.005	0.005	1.721
Copper	(Cu)	0.05	0.08	0.10
Zinc	(Zn)	1.0	0.0	2.4
Manganese	(Mn)	–	0.0	0.30
Aluminum	(Al)	–	0.18	20.6
Barium	(Ba)	–	0.0	5.0

[a]State of Florida quality standard for incinerator effluents. Data from Broward County, Florida incinerator. All values are expressed in mg/l.

Table 72. Effect of pH on Cation Concentration-Flyash Water, Incinerator A.[14]

	Concentration (ppm)	
Cations	pH = 3.4	pH = 6.30
Ca	913	718
Na	1621	1621
K	212	173
Mg	83	64
Zn	78	52
Pb	19	10
Al	32	14
Mn	4	3
Sr	1	1

Although good incinerator performance results in high burnout of organics and thus lower BOD in quench, scrubber, and other process waters, some BOD content of untreated waters is to be expected. Some bacterial content can also be expected as shown by a limited study of incinerator wastewaters from a 1.9 metric ton per hour (50 tons/day) batch feed incinerator and from an 11.3 metric ton per hour (300 tons/day) continuous feed municipal incinerator.[15]

The BOD and bacterial contents of the wastewater provide a potential for odor problems, especially where wastewaters are impounded and not sent directly to treatment plants. Although not commonly necessary, chlorine,[16] ozone,[17] quaternary ammonium phenate,[16] and other chemicals can be used for odor control. The quantity of such chemicals required would be greatly reduced by pretreatment, *e.g.*, by biological treatment.

Wastewater Treatment

There is an increasing trend in the design of process water systems to recycle at least a portion of the water to satisfy other process water requirements within the plant. Minimum wastewater treatment systems recommended for solid waste processing plants include settling basins or lagoons and pH adjustment. Oil skimmers and retention baffles may be used in the basin to handle leakage or spillage from machinery lubricators, the machine shop, and hydraulic systems.

Automatic pH control systems with a measurement probe and electronic controls which proportion the feeding of chemicals are useful. Chemicals used to adjust acidic wastewater include lime, soda ash, and caustic soda. Sulfuric acid is generally used to reduce pH when the wastewater is alkaline. As noted in the section on "Wastewater Quality," scrubber water is usually acidic, while overall wastewater may be acidic or basic.

Consideration should be given to reusing water which has been thus treated to reduce the fresh makeup water required and to minimize the quantity of contaminated water discharged. The quality of treated water from the settling basin will often be adequate to permit reuse in flue gas scrubbing (after filtering to prevent nozzle plugging), residue quenching, ash conveying, and for utility water used for washdown, etc. Sufficient fresh makeup water may be needed to prevent precipitation of scale in the piping and other water-handling equipment. A schematic drawing of a simple treatment system is shown in Figure 39.

It is desirable to be able to discharge wastewater to existing municipal treatment plants, but, where this is not possible, maximum recycle and a greater degree of on-site treatment should be practiced. A wastewater treatment system which goes beyond simple settling and neutralization may include flocculation, biological treatment, and filtration. These treatment steps, which will be outlined here, are used for many industrial plants though they have not been required for municipal incinerators.

Chemical flocculation involves the use of coagulants and coagulant aids to remove inorganic and organic contaminants. Alum is frequently

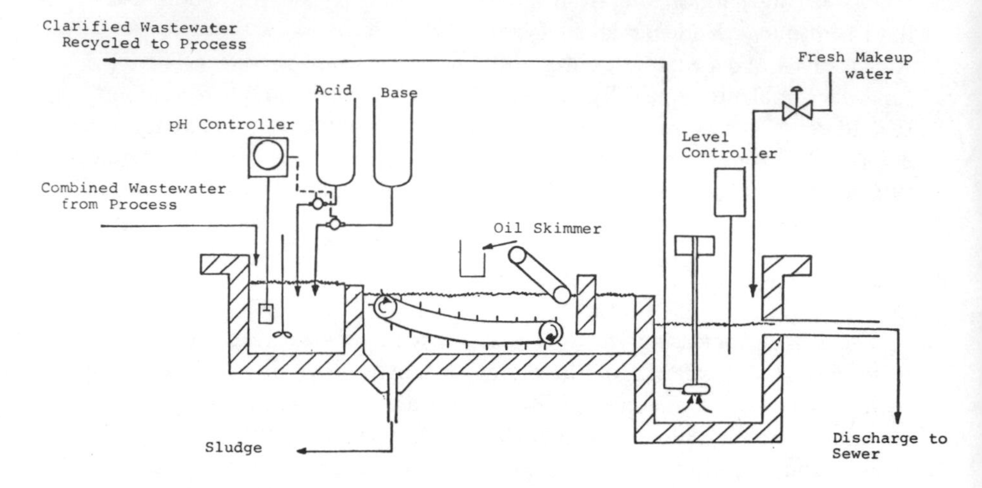

Figure 39. Wastewater treatment system for wastewater discharged to a sewage treatment plant.

used as a coagulant, and lime may be added to produce dense floc, which settles readily, and to simultaneously increase pH. Coagulant aids include polyelectrolytes and activated alumina. Suspended solids can be reduced to 20 mg/l in a well-designed, carefully-operated chemical treatment system. Dissolved solids concentrations will also be reduced substantially, particularly when calcium, magnesium, manganese, and iron are initially present in high concentrations. Chemically-aided flocculation is also employed as a means of reducing the concentration of heavy metals such as lead, chromium, copper, zinc, aluminum, barium, lead, manganese, and mercury.

Biological treatment depends upon contacting the wastewater with bacteria and other organisms to effect a metabolic breakdown of the organic substances present in the water. This can be achieved in various types of equipment, with the activated sludge process and the trickling filter being the most common. In activated sludge treatment, the organisms are kept in suspension throughout the wastewater by means of injected air or mechanical turbulence. The trickling filter, by contrast, contains a "fill" or solid matrix to which the organisms are attached and over which the water flows.

The process of choice depends upon capital and operating cost, strength of the waste to be treated, degree of purification required, availability of space, and other factors. Toxic materials including phenols, cyanide, and pesticides can be removed to varying degrees by biological treatment.

However, the design must provide positive means of control, such as equilization tanks, to insure that the concentration of these materials entering the unit are kept low enough to prevent upset to the organisms.

If biological treatment were contemplated, segregation of incinerator wastewater according to the degree of organic contamination present could be considered. This would permit biological treatment of residue quench water with simpler treatment of water from scrubbers, other air pollution control equipment, and flyash handling where BOD contamination is low.

Filtration of water by means of granular beds of sand or anthracite is sometimes used as the final step in the removal of suspended matter from water, usually following a settling step. Granular bed-type filters are used to polish the effluent from physical/chemical processes and/or biological treatment plants. Removal of 95% of the influent suspended solids and 30% of the BOD is possible. Addition of chemicals such as clay has been found useful in improving the removal efficiencies of granular bed filters.

Finally, disinfection might be required to allow discharge of the treated wastewater directly to the receiving water. Chlorine has been most commonly used, added in sufficient quantity to leave a 1 or 2 mg/l residual. However, toxic chlorinated organic compounds can be a problem, casting some doubt upon the desirability of chlorine addition. The use of ozone as a disinfectant has been the subject of recent investigations, primarily because toxic by-products do not appear to be generated. Also, ozone is more effective in removing traces of cyanides and phenol from the water.

Figure 40 is a schematic diagram showing the type of treatment system which might be used where high contaminant removal efficiencies are required. It should be noted that this more extensive treatment results in sludges which must be disposed of, just as is the case with sewage treatment plants. In general, where possible, it is most desirable to maximize recycle after simple on-site wastewater treatment, discharging effluents to the municipal sewer system.

DISCHARGES FROM PYROLYSIS PROCESSES

In recent years, thermal processing techniques have been developed which produce fuels by the pyrolysis of solid waste in an oxygen-deficient or -reducing atmosphere. In some pyrolysis processes, including the current versions of the Monsanto Process (Baltimore, Maryland plant) and the Torrax Process (Orchard Park, New York pilot plant), the hot fuel

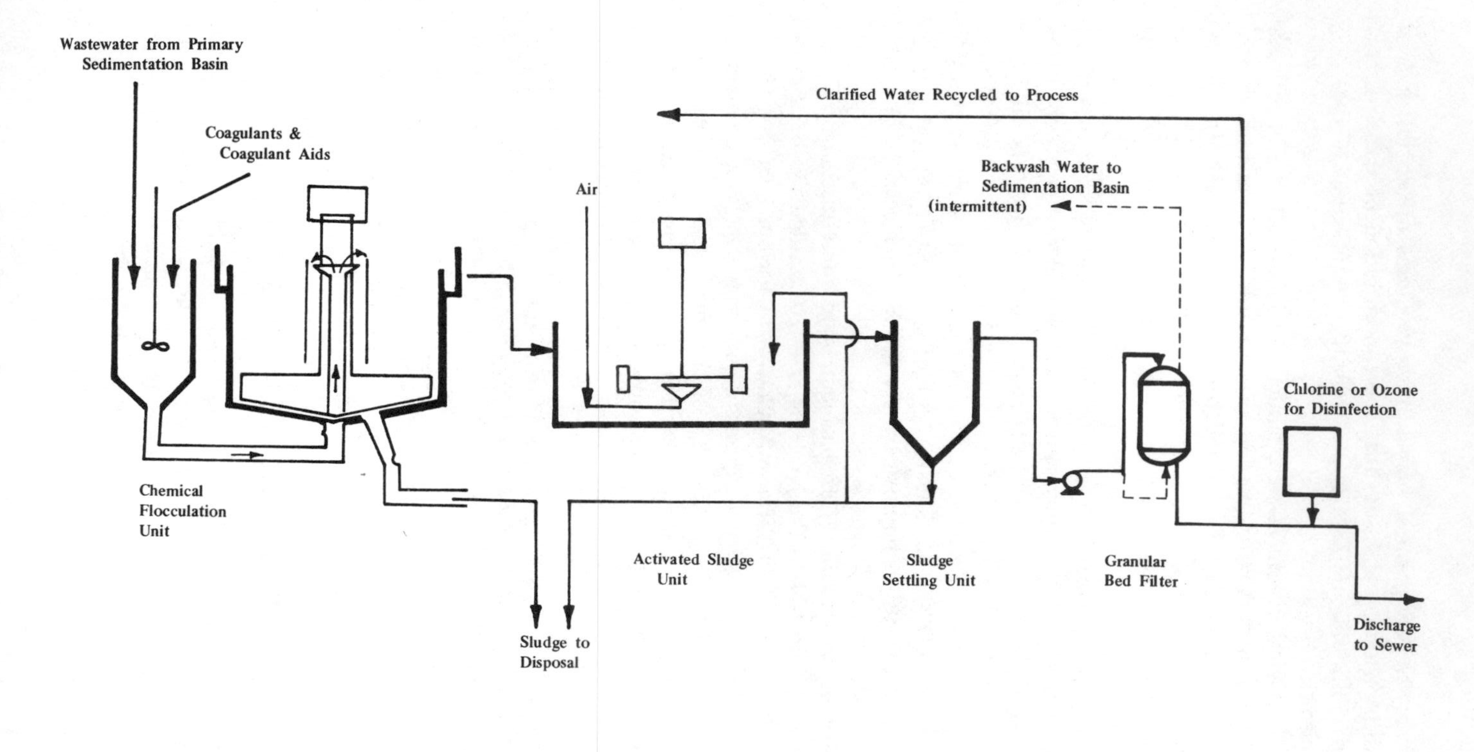

Figure 40. High-efficiency wastewater treatment system.

gases are burned in a waste heat boiler to generate steam. After combustion, the waste gases are scrubbed with water before being released to the atmosphere. Control of air emissions and treatment of scrubber water pose no special problems which are not encountered in the operation of conventional incinerator plants.

Other processes, including the Occidental Research Company Process (San Diego County, California plant), condense the pyrolytic fuel oil produced. The condensate contains both the liquid oil and the water which is produced simultaneously during pyrolysis. These separate into an oil phase and a water phase which are physically separated. However, the water phase is highly contaminated with a multitude of water-soluble organic compounds such as acids, aldehydes, and alcohols. This contaminated water, which may contain BOD values near 100,000 mg/l, poses serious recovery or disposal problems.

In Occidental's San Diego plant, which is under construction, the aqueous phase will be vaporized and the organic contaminants burned with a net consumption of energy. A second alternative which has been considered is to blend the aqueous condensate with other wastewater and to subject the combined waste to biological treatment. Recovery of the organic compounds from the water phase may be possible, but considerable development work would be required to find a practical and economical method for recovery.

REFERENCES

1. DeMarco, J. *et al.* Municipal-Scale Incinerator Design and Operation. PHS Publication No. 2012. U.S. Government Printing Office, Washington, D. C. 1969. (formerly Incinerator Guidelines-1969).
2. Sundberg, R. and G. H. Weyermuller, Ed. Ozonator Operating Cost Only $2600/Yr. *Chemical Processing*, January 1970.
3. Roeder, W. F. Carbon Filters Control Odors at Refuse Transfer Station. *Public Works* **100**(4): 96-97. April 1969.
4. Kaiser, E. R., *et al.* Municipal Incinerator Refuse and Residue. Proceedings, 1968 National Incinerator Conference (New York, May 5-8, 1968). American Society of Mechanical Engineers, pages 147-152.
5. Fernandes, J. H. Incinerator Air Pollution Control. Proceedings, 1968 National Incinerator Conference (New York, May 5-8, 1968). American Society of Mechanical Engineers, page 102.
6. Thermal Processing and Land Disposal of Solid Waste. U.S. Environmental Protection Agency. *Federal Register* 39(158) Part III: 29328-29338. August 14, 1974.
7. Hagerty, J. S. *et al.* *Solid Waste Management.* Van Nostrand Reinhold Company, New York. 1973. 302 pages.

8. Bowen, I. G. and L. Brealey. Incinerator Ash–Criteria of Performance. Proceedings, 1968 National Incinerator Conference (New York, May 5-8, 1968). American Society of Mechanical Engineers, pages 18-22.
9. Schoenberger, R. J. and P. W. Purdom. Classification of Incinerator Residue. Proceedings, 1968 National Incinerator Conference (New York, May 5-8, 1968). American Society of Mechanical Engineers, pages 237-241.
10. Eberhardt, H. and W. Mayer. Experience with Refuse Incinerators in Europe. Proceedings, 1968 National Incinerator Conference (New York, May 5-8, 1968). American Society of Mechanical Engineers, pages 76-77.
11. Achinger, W. C. and L. E. Daniels. Seven Incinerators. SW-51 ts.lj. U.S. Environmental Protection Agency. 1970. 64 pages.
12. Jens, W. and F. R. Rehm. Municipal Incineration and Air Pollution Control. Proceedings, 1966 National Incinerator Conference (New York, May 1-4, 1966). American Society of Mechanical Engineers, pages 74-83.
13. Schoenberger, R. J. and P. W. Purdom. Characterization and Treatment of Incinerator Process Waters. Proceedings, 1970 National Incinerator Conference (New York, May 17-20, 1970). American Society of Mechanical Engineers, page 206.
14. Wilson, D. A. and R. E. Brown. Characterization of Several Incinerator Process Waters. Proceedings, 1970 National Incinerator Conference (New York, May 17-20, 1970). American Society of Mechanical Engineers, page 199.
15. Tucker, M. G. Biological Characteristics of Incinerator Wastewaters. Unpublished graduate student research project in CE 687 course. University of Michigan, August 1967, 15 pages.
16. Matusky, F. E. and R. K. Hampton. Incinerator Wastewater. Proceedings, 1968 National Incinerator Conference (New York, May 5-8, 1968). American Society of Mechanical Engineers, pages 201-203.
17. Ozone Water Treatment Nears Pilot Stage. *Chemical Engineering News.* September 8, 1969.

INDEX

INDEX